NICHING UP

THE NARROWER
THE MARKET,
THE BIGGER THE PRIZE

NICHING UP

CHRIS DREYER

COPYRIGHT © 2022 CHRIS DREYER
All rights reserved.

NICHING UP
The Narrower the Market, the Bigger the Prize

FIRST EDITION

ISBN 978-1-5445-3243-1 *Hardcover*
      978-1-5445-3242-4 *Paperback*
      978-1-5445-3244-8 *Ebook*
      978-1-5445-3245-5 *Audiobook*

FOR GREY,

I LOVE YOU, SON.

CONTENTS

INTRODUCTION 9

1. ATTACK OF THE CONS 17
2. AWARENESS 33
3. EXPERTISE 51
4. PREMIUM PRICING 65
5. CONVERSIONS 75
6. RELATIONSHIP EQUITY 87
7. REFERRALS 103
8. EFFICIENCY 119

CONCLUSION 129

ACKNOWLEDGMENTS 141

ABOUT THE AUTHOR 143

INTRODUCTION

Anything I've been successful at can be attributed, in some way, to niching.

In 2006, I started doing affiliate marketing as a side hustle from my full-time job as a basketball coach and detention supervisor. The very first website I made was loseadoublechin.com. It started out as sort of a joke, but the reality behind it is that I was overwhelmed by the thought of tackling the diet and weight loss industry. I didn't feel secure in my knowledge, not being a physician or having a medical background, or about learning such a broad and complicated topic. On the other hand, chins are just a small part of the body, and I thought I could learn enough about this one area by reading as much information as possible.

So I created this site and I soon ranked number one for "double chin" on Google. It was a terrible site, but it ranked number one for several years. It was quite successful, and it made a lot of money.

Then I started thinking, "This one site did so well and it's not even my best work. I bet I can do this again!"

I ended up making eighty other websites. I earned a hundred dollars here, two hundred bucks there, but nothing like the success of the double chin site. My first mistake—reflecting on this many years in the future—was that I didn't appreciate what I had done right. I had niched!

I should have realized what worked and run with it.

When I decided to start my digital marketing agency, I initially launched attorneyrankings.org (though we are now rankings.io). I knew I wanted to work in the legal industry, but I was scared that every time I gave up a service, it would hurt my revenue or my positioning. In actuality, every time I chose to not do one thing, other doors opened. I discovered a propensity for search engine optimization (SEO), and I started to find more success.

When I went to Vistage (which is a peer advisory group for CEOs) and had my first business review, I was told, "You need to offer your services to physicians, home services, and these other niches, because you're doing so well with legal."

I didn't take my own advice or learn from my previous experience with all those affiliate sites. Instead I listened to my peers (who, in retrospect, were in industries that were completely different from mine) and went after a larger cap.

I signed a dental practice and a few other clients in other niches—and my momentum slowed. I had more people to sell to, but I was growing at a slower pace. If you've seen those stock market charts with the zig-zag arrow showing ups and downs, this was a big red arrow pointing downward. I kept asking myself, "What's going on?"

As you've probably guessed, I had severely discounted my positioning as an expert when I was only specializing in legal.

Thankfully I didn't stay on this path too long. We fixed our positioning, and the arrow started pointing upward again.

This time I learned—so well, in fact, that niching as a provider of SEO specifically to personal injury attorneys has contributed to our company being an Inc 5000 recipient for the past five years in a row.

I also learned that anything I've ever been successful at is because I've been obsessively focused. I won fourth place in a 14,000-player poker tournament because I focused on nothing else for a whole month and reviewed 40,000 hands of poker during that time. I was a top-rated collectible card player because there were many clans to play, but I learned the ins and outs of just a few. In sports, I was successful because I focused on basketball, went to summer basketball camp, and put all my hours and attention into training for just basketball. That focus led to me becoming captain of our conference and being offered scholarships to play college basketball.

I'll tell these stories in more detail later, but for now suffice it to say that I owe my success to *niching up*.

DOESN'T NICHING MEAN SAYING NO?

Every time I've considered niching up, it feels like stepping off a cliff. It's scary at first—*what if I don't know what I'm doing? What if I fail?* But every time I've taken that step, it ends up being a really positive decision.

Why? Because niching up means moving forward from a place of abundance and growth.

The first thing many people think of when they hear niching is scarcity; they think of having fewer opportunities, of less ability to monetize or grow their business, of *taking something away*. Many individuals' biggest fear with niching is that by saying yes to one area, they're saying no to many others. It's true: niching does mean saying no to business and shrinking your market. For example, for me to become a standout basketball player, I had to focus on that one sport—which meant I had to quit practicing baseball as much.

But in reality, niching provides opportunities and gives you *optionality*. Even after you choose a niche, if you get a lead that's not where your deep expertise lies, you don't have to say no. In many cases you will, because it's not your main focus, but you always have the option of saying yes if it's an opportunity you particularly want to explore. In some cases, you can genuinely help with something that's not specific to your niche, so you don't have to look at it only as black or white, yes or no.

By saying no to some things, niching opens the door to many things—including more or better possibilities.

And that knowledge—that you don't have to slam the door shut on other business opportunities just because you choose to niche up—takes away some of that fear. I find it reassuring to know that just because I choose to focus on helping personal injury attorneys with SEO, that doesn't mean no other opportunities can ever come through my door. In reality, I do mostly personal injury (PI) but I have the option of saying yes to something else if it makes sense for me.

In fact, of my forty-five current clients, I have three that are not personal injury attorneys. Three may not be a lot, but that's three times I said yes outside my niche.

Most recently, I had a divorce attorney come to me and say, "Let's rock and roll!"

I knew I could serve and help them—their keywords and SEO strategy are similar, and they had the right attitude—so I said yes. But that doesn't mean that I'm going to now market myself for both PI *and* divorce law SEO.

Far from narrowing your options, niching up opens a world of possibility to say yes to the people who are right for you and your business—the ones you are best suited to help.

RICHES ARE IN THE NICHES

Telling stories has always been my go-to method for teaching.

I've spun tales for high schoolers about how an underdog hoops upstart was able to join the first string through good choices and hard work. I've doled out yarns to website visitors about extreme double-chin reduction achieved through clean eating and exercise regimes. And, through the stories in this book, I hope I am able to help you learn from my journey with niching—both from my mistakes and what I got right.

Along the way, I'll help you to see your own existing business, or the business you dream of founding, through the niching lens. You'll understand the many benefits niching can offer, and how it can be easily applied to any business vertical. You'll learn how to figure out if niching is right for you simply by determining who it is that you are able to serve best and most efficiently. Finally, I'll show you that you don't need to undertake a ton of research or spend money to do any of it. You already collect all of the data that you'll need.

When you choose a specialization, your focus and attention make all the difference. Niching allows you to stand out, to be in a blue ocean instead of a red one. And when you niche up, you obtain significant advantages.

That doesn't mean that there are no downsides, just that everything you gain through niching outweighs those disadvantages. In fact, we'll address the potential cons of niching in Chapter 1, and the rest of the chapters will show you all the benefits of niching up—things like:

- having greater *awareness* of the opportunities open to you, so you can choose where to say yes or no;
- gaining *expertise* through hours of practice, which sets you apart in your field;
- being able to charge *premium prices* because people feel you've invested more time in becoming an expert and therefore they're willing to pay more for it;

- more easily *converting* prospects to clients because you better understand your niche audience;
- having better *relationships* with those in your industry, which leads to more goodwill;
- getting—and giving—more *referrals*, which means helping more people;
- and becoming more *efficient* as you develop repeatable processes designed to help you and your clients.

Let me be clear that niching is not for everyone and this book is not my attempt to convince you that every business ought to serve a niche audience. If you're a business owner who is happy with your margins and the manner in which you're competing, perhaps you'll read this book and still decide to stick with what's working. I've seen niching make a positive difference for entrepreneurs who sell everything from Facebook ads to baskets of fries, but I'm not arrogant enough to believe that I know what's best for every business. I'm just sharing my experience with niching, how it's benefited me, and how I think it may be able to benefit you, too.

This book is also not a step-by-step guide to finding your specific niche. Every niching decision is unique, so while I can absolutely give you a primer on the facts and figures you should pull together in order to figure it out, I can't make the determination for you. By the time you're finished reading you'll be armed with all you need to reach your own conclusions.

I'm excited to invite you along on my journey, and I can't wait for you to understand why I believe so deeply that, for most businesses, the niches hold your riches.

CHAPTER 1

↓

ATTACK OF THE CONS

The first time I watched *Star Wars: Episode II - Attack of the Clones*, I immediately noticed something different about one of the lightsabers in the Battle of Geonosis. In the scene, a bunch of random Jedi warriors are fighting, holding lightsabers with the familiar blue and green colors any *Star Wars* fan would expect.

But somewhere deep in the mix, I caught sight of something different and did a double take. A purple band of light came in and out of view, and I said, "That's badass! Who's mowing down battle droids with a purple lightsaber?"

The Jedi on the other end of that purple lightsaber was, of course, Mace Windu, played by the incomparable Samuel L. Jackson. It turns out that his purple lightsaber was designed by Jackson to elicit precisely the reaction I experienced. The idea of having Mace fight using a purple lightsaber was, in fact, the actor's own idea.

Several magazines quote Jackson as saying, "We had this big arena, this fight scene with all these Jedi and they're fightin' or whatever. And I was like, well shit, I wanna be able to find myself in this big ol' scene. So I said to George, "You think maybe I can get a purple lightsaber?"'"

He just...asked! Who knows if this tactic would have worked for a lesser actor. I mean, can you picture some unknown guy asking George Lucas himself for special treatment so he could stand out more in the next *Star Wars* movie? It worked because this wasn't just any actor asking for the lightsaber upgrade; it was one of the most well-known names in film, who simply wanted to make sure he (and his fans) could find his recognizable face on the big screen!

So Jackson got his custom-colored lightsaber, and fans got the perfect way to pick him out of the crowd in any fight scene. You might not be able to differentiate the other Jedis from one another if you can't see their faces on the battlefield, but thanks to that purple lightsaber, you can always pick out Mace, even from a distance.

NICHING IS NOT JUST FOR NERDS

What on earth does this *Star Wars* anecdote have to do with marketing? Well, Mace's purple lightsaber perfectly proves a fundamental truth of niching: *when you do something that no one else is doing, you stand out.*

> When you're a *Star Wars* nerd obsessed with niching, these are the kinds of things that run through your brain. Give me the space to daydream, and it's only a matter of time before I'm thinking about *Star Wars*...and probably also relating it to marketing.

As you know (especially if you're on the nerdy side, like me), standing out can be great—but it has its downsides as well. Every coach and entrepreneur loves to refer to niching as the Holy Grail of owning a business, and they always talk about the pros, but they tend to skip over the cons.

I understand the instinct here. When you're selling someone on a new concept, generally you want to highlight all the great things about it. But the reality is that there are both advantages and disadvantages to niching. The pros far outweigh the cons, but those cons *do* exist and we need to address them.

Why?

Because I want to guide you in the right direction. I don't want you to only see the advantages, dive into a niche, and then discover that it won't be beneficial to you. It is a disservice to pretend that everything is sunshine and rainbows when the truth is that there are a lot of sunny days when you niche up… but it can also be dark and cloudy. It doesn't help anyone to sweep these very real issues under the rug.

There's also not necessarily a solution for any of these cons; they are just potential problems you need to be aware of *before* you decide to niche—and definitely before you decide which niche is right for you. As I said, there are more advantages than disadvantages, which makes up for any drawbacks.

But forewarned is forearmed, as they say, so grab your lightsaber and let's get realistic about niching. It's time to take a closer look at the cons.

CON #1: SMALLER MARKET

The first con that I think everyone is aware of is that when you niche there is a smaller market and, therefore, fewer buyers. The very definition of niche means it is limited to a specialized audience. Depending on the specific business niche, the ability to target customers or audiences can be constrained, which affects business growth.

Simply put, serving a niche means fewer customers.

When you're going after a smaller niche, you're giving up the larger share of the market. The biggest associated risk is that if there aren't enough interested buyers, then there's no profit to be made. You can get reports and check surveys of how many business owners are in that market—but ultimately, at some point, you are going to exhaust your market.

Once you tap your market, your options are either to create additional products and services, or to find additional customers. Let me give you an example of what this looks like, using my company. We are very well known for performing SEO services (a niche within the marketing industry) for personal injury attorneys (a niche within the legal industry). If we have clients in most of the major metros, we have reached the extent of our market. Once that happens, to grow the business any further we either have to sell other services or open our existing service up to other areas of the law—other niches.

There can be a flipside to this as well: some customers don't recognize the advantages of niching and are hesitant to work with someone who niches, thus making an already small market even *smaller*. For example, sometimes I will talk to a law firm that does personal injury, criminal defense, and bankruptcy law (so clearly they do *not* niche), and they don't want to work with my agency because we focus on PI. Because they choose not to niche, they don't see the value of working with someone who is an expert in their own field. The lesson here is that not only are you niching, you're also looking for other people who are doing the same in their industry. Just as "game recognize game," expertise also recognizes expertise.

CON #2: WASTE

The second con (which may not be as immediately obvious as the first) is that it can be more difficult and cost you more money to get in front of your target audience when dealing with a niche. It's difficult to target sub-niches with ads and other traditional forms of marketing.

If I want to sell SEO to personal injury attorneys, I can't go through traditional marketing channels to try to rank for personal injury SEO because it just doesn't get typed into Google very frequently. I can't go to Google ads, bid on relevant phrases, and expect a large pipeline of leads (even though there are about 93,000 personal injury law firms in the US,[1] those search terms only get entered about ten times a month).

As much as Facebook and Google ads have advanced from a targeting perspective, they still don't have enough information to be as specific as you'd need for many niches.

---

[1] It is estimated that there were approximately 92,900 personal injury lawyers in the United States in 2021 (somewhere between 5 percent to 7 percent of all lawyers in the United States). https://adidemlaw.com/blog/2019/08/24/size-of-personal-injury-legal-market/

This is especially true for the topics personal injury attorneys use to niche up, like truck accidents or mass torts. If Joe Fried (a personal injury attorney at Fried-Goldberg LLC who specializes in truck accidents) bids on the search term "motor vehicle accidents" (because it gets typed in more than "truck accidents"), his results are also going to show people searching for information about motorcycle and car accidents, mixed in with the truck accidents he's actually searching for.

I also can't do very specific types of marketing, like running a radio ad campaign, because how can I reach all personal injury attorneys? I could buy an ad on sports radio, for example, but are enough PI attorneys going to listen to that station to make the buy worth it? Probably not. There are paid legal directories I can advertise in, but they don't have a directory strictly for personal injury; there's just not enough volume.

Instead, niching requires *relationship building,* because you can't advertise with direct marketing like you could in broader industries. I go to conferences where PI attorneys congregate so I can interact with them in their specific communities. Similarly, Joe Fried has become a thought leader on truck accidents. (We'll look at this concept of relationship equity more in Chapter 6.)

Because your message is so specific, you also have to spend more to acquire clients, whereas if you had that broader audience, the same spend would get you more clients. It's not as straightforward as for a personal injury attorney who might put a billboard on the road and run a radio ad saying, "Hey, I'm a personal injury attorney. If you've been hurt, contact me," and effectively reach a whole bunch of people who potentially need their services.

> Now, in many cases those customers you've spent more to attract are actually worth more because you can charge higher fees (or, in the case of a personal injury attorney, you can earn higher contingency settlements)—along with all the other advantages that come with niching, as you'll see in the chapters that follow.

CON #3: COMPETITION

Niching also introduces competition. You may be the first (or one of the first) in your niche, but once you experience some success and people see you're doing well, they might start to think, "Hey, I can do something like that too!"

Look no further than the history of the motor vehicle industry for evidence of this. Before Henry Ford, no one had cars—so he had a bunch of people to sell to. Once everyone had that first car, he couldn't just sell them another Model T; he had to create a different type of car. Remember what I said in Con #1? Once you've tapped your market, you need either more customers or more products/services.

Here's the thing: Ford found a really good niche—say, oh, being the very first car manufacturer—and that invited competition. Regardless of your niche, once you've identified it, others *will* follow suit because of your success; they'll be right behind you, on the trail that you blazed. Henry Ford would have loved to just continue making Model Ts, but he couldn't because Chevrolet and other car makers started coming in. He had to make different cars.

As you can see, these cons can be layered. You can have both a smaller market share and invite competition; it's not just one or the other.

Joe Fried started out as the only personal injury attorney specializing in trucking accidents, and people thought he was crazy. Once he started crushing it—today he's the leading truck accident attorney nationwide—people came out of the woodwork to follow his lead. They just let him figure it out first.

Similarly, Steven Levin (of Levin & Perconti in Chicago) was one of the first people who created the niche of nursing home neglect cases. Now he has legal precedent—and a bunch of competition after his discovery of a good niche.

> Inviting increased competition can certainly be one of the disadvantages of niching—but like most of these cons, there can also be an upside to it. In the book *The 22 Immutable Laws of Marketing*, authors Al Ries and Jack Trout highlight that one of those immutable laws is being first in a market. We don't talk about who made the second car; we talk about Henry Ford being first. There's definitely an advantage to being number one in a particular niche, from a perception perspective.

The other side of this con is that at the very beginning of creating your niche, you *don't* have competition—and having those competitors nipping at your heels, always trying to take your customers, is what makes you better and drives you to grow and innovate.

Henry Ford's Model T was atrocious looking, but that didn't matter when it was the only car that existed. As soon as other companies came along and started making nicer-looking cars like Corvettes, he couldn't get away with just cranking out boxes with wheels. Had he never had any competition in his niche, we would probably be driving something much different—and definitely not as fast.

Similarly, Elon Musk got everybody talking about electric vehicles with Tesla; once he actually made it happen, Volkswagen, Subaru, and everybody else moved to create an electric option. In this way, competition can be an advantage as much as a disadvantage.

CON #4: LACK OF DIVERSITY

Once you decide on a niche, there is a lack of diversity in your work. You are basically doing the same thing and talking to the same people all the time. If you don't have a passion for what you're doing, it can get monotonous.

Let's use a sports example (see, I'm not just a nerd!) and imagine that you had to play basketball all day, every day. Maybe you get burned out, or over time you just feel like life is passing you by. There's no creativity, no learning, no new challenges, and a lack of new experiences. This probably sounds very boring to someone who doesn't play basketball—but the experience is very different for a serious basketball player who feels passionate about the game. Their coaches are breaking down their game, helping them work on different shots, and encouraging them to practice how fast they can move down the court. They're not just playing the same game all day, every day—and you're probably not actually doing the same thing all the time either.

That's not to say that even professionals can't get sick of their niche. Just think about Michael Jordan, who played basketball and became arguably one of the—if not *the*—greatest players of all time. There was still a period where he went to play major league baseball. Perhaps he got bored with basketball and wanted to play baseball for a new challenge, to learn new skills, and to discover a new frontier.

Ultimately, however, it didn't take (you'll note that we don't talk about Michael Jordan as a great baseball player). He's a good example of why this lack of diversity in your day-to-day can be a downside of niching—and that, for some of these cons, there's no solution. It just is what it is, and it bears acknowledging.

I haven't experienced this personally, because I have a super competitive personality, but some creative types might find that doing the same thing feels monotonous. Whatever your personality, if you're not passionate about what you do, you're probably not going to put out a good product—but you're also not likely to consider that niche in the first place.

If the cons in this first chapter talk you out of niching, maybe you don't have the necessary passion for that niche. You need to have passion before you can go in and potentially spend more money to acquire fewer clients. There has to be something there for you that makes all of this worth it.

CON #5: INDUSTRY RISK

Industry risk is another easily envisioned con of niching. Imagine that you were in the cruise line industry when the COVID-19 pandemic hit. All of a sudden, ships were considered petri dishes of diseases and the prevailing attitude was, "Who would ever want to go on cruises??" It didn't matter if you had the best cruise line, with a wicked slide, the best buffet, and an amazing onboard magician—when COVID hit, people were not going on cruises.

Every industry has some level of risk. Many personal injury attorneys deal with a large number of motor vehicle accidents.

Well, at the beginning of COVID, more people were working from home, which meant fewer people driving, which meant far fewer auto accidents—with the accompanying cases and settlements. I'm not a personal injury law firm, but because I sell to them, I was also impacted—when they have fewer accident cases coming in, they have less money to spend on marketing.

Restaurants closed or were forced to change how they operated during the first waves of the pandemic. With that, however, came accelerated technology and forced innovation. Suddenly, Grubhub, DoorDash, and Uber Eats were the lifeline between those restaurants getting business and you not having to decide what to make for dinner again.

Attorneys who previously had to show up in court for everything were allowed to start making appearances via Zoom.

Technological innovations like the ones we saw take off during the pandemic are a double-edged sword because they quickly render aspects of their industries outdated. If you own a bunch of taxi cabs, Uber and Lyft didn't do you any favors. Similarly, those of us of a certain age watched as VHS gave way to LaserDiscs (remember those? I'm dating myself), then to DVDs, and now to streaming. Even mail service is basically outdated because we now do so much digitally.

As you can see, there can be technological risks in choosing a niche. Once we get flying cars, who knows if we'll even need attorneys specializing in auto accidents (I never saw George Jetson and his contemporaries running into each other in the sky). Then, with those innovations, will come a whole new set of problems that we can't even anticipate until they're more of a reality. Some niches will be eliminated with the new advancements—but somebody else will need to niche into solving those problems.

> Again, every industry has its risk—but every risk means that industry is ripe for innovation and disruption. Is that a pro or a con? You decide.

CON #6: PRODUCT PERFECTION

Another potential con of niching is product perfection. Competition in the marketplace requires companies to develop perfect strategies and provide better solutions. Because there are fewer prospects when you niche, there is less room for error. You must be able to provide products (or services) that are exactly in line with what the customer demands.

What this means is you get fewer reps at the plate. You only have so many buyers in your niche, so you can't turn out a poor product because you won't have the opportunity to make up those sales elsewhere. If you're going to call yourself an expert (and that's what you're doing when you niche up), you lose all integrity and credibility with a buyer if you show yourself not to be that expert they expect. Every sales conversation is based upon trust, and the moment they lose trust in you, that sale is lost and your future sales are greatly diminished.

If you take a swing and miss, you risk damaging your reputation in a small market, which means that future opportunities to sell something to those individuals—even if it's a different, better product—will be unlikely. You have to deliver your best at every opportunity, even if you're just getting started. You have to go out there and hit home runs, even when you're just getting into that niche and may not have everything figured out yet.

Here's the kicker: this is true every single time you have a new product or service. Even if you successfully sell your first product to everyone in your market, this disadvantage rears its ugly head with the *next* product. You will have a brand-new product—and a brand-new opportunity to go down in flames.

CON #7: INCREASED EFFORT AND SACRIFICE FOR BUYERS

The final con on our list may be the biggest of all: when you niche, you may make things more difficult for *your buyers* by increasing the amount of effort and sacrifice they have to make. Let's say, for example, that John needs to purchase a watch and a pair of shoes. If he goes to a niched store like Rolex, which *only* sells watches, John now has to sacrifice more of his time and effort to go to *another* store to buy his shoes. It's far more convenient to just go to one store, like Walmart, that sells both.

As another example, if a client of ours wants a full digital marketing campaign, they have to use multiple vendors because we niched up to only doing SEO. So they have more meetings, more points of contact, etc.—hence raising the effort.

In fact, this is the exact reason Walmart introduced selling groceries at their stores. Sam Walton didn't want to sell produce, but he knew his buyers would like the ease of being able to buy home goods *and* their groceries in one trip. It's easier to go somewhere that has everything in one place rather than going to multiple stores...even if those multiple stores may have better individual goods. A buyer is likely to get significantly more value from making two separate purchases at niched stores, but they may choose to sacrifice some of that quality for the convenience of buying both at once.[2]

BUT IT'S NOT ALL BAD NEWS!

When you look at the negative side of niching, you'll see that many of these cons are related. When you niche and you don't have competition—when you're the only one in a market—your product may not improve because competition is what forces you to improve your product. If you have a super small niche—let's say ten buyers, to make it an extreme example—once you sell your product to one of the ten, if it's not good, you've only got nine more chances...and your reputation may already be tarnished to those other nine. You can't just fake it till you make it in a niche; you have to be an expert because you have limited opportunities.

To the best of my knowledge, my agency was the first to specialize in *only* personal injury attorneys, and I had no competitors of note at the very beginning. Our success has caused other agencies to follow not only our positioning, but also our focus. Even though they're not copying the niche exactly, they've seen the benefits of our approach and are modeling many of our behaviors.

Different industries and different markets are going to experience these cons at differing levels. I can't tell you the specifics

---

[2] You can learn more about The Value Equation, which includes Effort & Sacrifice as one factor, in the book *$100M Offers: How To Make Offers So Good People Feel Stupid Saying No* by Alex Hormozi.

of your market, your niche, or your subset. I can tell you that, generally speaking, these are the cons that everybody is going to experience at some level.

The smaller your market gets, the faster these cons appear. If you're going to sell SEO in the legal industry, you have tons of law firms to market yourself to…but once you niche into PI, you have fewer firms. If you then sub-niche into, say, mass torts (a subset of PI), you could tap that market even faster.

It's important to note that the deeper you go in a niche, the faster these cons are accelerated—and the greater the risks can impact you.

I don't want to scare you away from niching—but neither do I want you to wander in with blinders on. You now know the potential disadvantages of niching, and the remaining chapters will focus on the advantages to niching up and how to use those advantages, well, to your advantage.

In the next chapter, I'm going to go against a piece of popular advice. Many people (particularly coaches and mentors) will tell you to just jump straight into your niche, but I think in most cases you should go broader before niching up. Going broad gives you an ability to see opportunities that you might otherwise have missed. It's that *awareness* that we'll focus on in Chapter 2.

CHAPTER 2
↓
AWARENESS

After I graduated college, I became a teacher and, to actually make some money (sheesh, I was broke), I started doing affiliate marketing on the side. In my first year, I probably only made a couple Benjamins...but by my second year, my side hustle was growing more profitable than my "actual" career (although, if you know anything about what teachers are paid, you know that's not a high hurdle to clear). I chose to focus those marketing efforts on health and fitness, and I niched into double chins, launching loseyourdoublechin.com largely as a bit of a joke, but also partly because I was overweight.

I fell into this niche naturally. I didn't even understand the pros and cons, but I *did* know that if I tried to make a website on health and wellness, I would be competing with everyone. I discovered, however, that there's not a lot out there about just the double chin aspect. I thought maybe there was something to becoming the "double chin guru."

Something you should know about me is that I get...let's call it "invested." People around me have been known to call me obsessive, but I like to think it's just that I'm super competitive. If I'm going to bother doing a thing, I'm going to be *good* at it. What constitutes "good" might vary, but at the very least, I want to be sure that I'm doing the best that *I* can do. That character trait reared its head as I started jumping into the double-chin business. I went into tunnel vision mode and read absolutely every word that I could find about double chins. Once I did that, I buckled down and tried to synthesize that information into what would be the most valuable to my reader.

I wrote all about what I had learned, then made sure that anyone searching Google for a double-chin cure would get served up my site. Soon people were coming to my site to read the well-optimized posts I had cooked up on their hideous turkey gobblers. I did my best to provide my readers with information that would help answer their questions; sometimes, that led to a commission through an affiliate product.

In most cases, to lose a double chin, you need to either lose weight or have surgery. You're probably thinking, "No shit, Chris." But what I also found, while diving into this niche, was that there was a massive audience that related to *hiding* a double chin.

Did you know that a large percentage of overweight males grow a beard to hide their double chin?

Little discoveries like that appealed to a segment of my audience. But I would have never uncovered this had I just focused on general health and weight loss content.

"UH, CHRIS, I HAVE A QUESTION"

When I talk about niching up, the two questions I get asked most often are "should I niche?" and "when should I niche?"

Let me answer the first question first, and I'll answer the second by the end of the chapter.

So the first question is, should you niche? The answer is... maybe.

In the beginning, before you decide to niche, you should start broader by gaining more experience in your desired industry. You may not know if niching offers the advantages that you need, or that there is a true opportunity until you really understand the industry or market that you're in. If you just decide to go narrow out of the gate, it could be difficult to find buyers—or to even know which is the right niche for you.

> David Epstein's book *Range: Why Generalists Triumph in a Specialized World* is a great resource that talks about how to be a generalist before you choose your expertise.

Before you niche, you need more experience so you can identify which niche offers the best opportunities for you. Spanish tennis star Rafael Nadal, for example, played multiple sports before identifying that he was better at tennis, and going all-in on that one sport (he's now one of the best players in the world).

When you go broad and gain more generalized experience, you will have more *awareness* about your industry and market, as well as about potential niches. I first got experience in the general legal field before I chose my niche. During that time, I didn't have the best experiences with criminal defense attorneys; quite frankly, many of them were just assholes (don't hate the player, hate the game). Though it's a fine niche profit-wise, I decided to swerve to the right due to *my* experiences.

If I had gone into the niche of working with criminal defense attorneys right away, I would have bypassed the PI niche I'm currently in (which I thoroughly enjoy) and ended up working with people I may not have connected with as well. Now, if I had gone all-in on that niche, it's quite likely that I would have found some criminal defense attorneys who *weren't* assholes (unicorns are rad pets too)…but maybe not—and then I would have been in a niche with people I didn't want to spend my time working with. That's one of the risks.

Look, I get it…criminal defense attorneys frequently have to defend their clients against serious charges (like, y'know, murder). They yell and scream to make their points to a jury, rather than exhibiting the calm demeanor typically implemented by PI attorneys to elicit empathy and get maximum compensation for a person who has been wronged. They tend to have completely different personalities, but if you talk to someone who has never worked in the legal field, they don't see those differences. An attorney's an attorney, right? They'd think that as long as you're working with any subgroup of attorneys, you're fine. These individuals don't have the awareness of the distinctions between the two niches.

> I think some people's negative associations with attorneys come largely from certain subsets of attorneys. If you're dealing with a bankruptcy or going through a divorce, you're probably not going to have a great experience with fond memories. Those specific negative associations elicit feelings and can give all attorneys a bad name because they all get painted with the same brush.

Another benefit of first looking at the broader picture before choosing a niche is the ability to discover whether or not you have a natural competency in that niche. You may see an opportunity for renovating and flipping houses, for example; a lot of people need that service, and there's a lot of money to be made. But even if you identify that there could be an opportunity there, if you are not a flipper (and not interested in becoming a flipper), you're not going to be successful. If you don't have both the passion *and* the natural aptitude in an area, you're not likely to excel.

Awareness is the first advantage of niching, so to get the most from it, we're going to look at getting experience so you know which niche makes the most sense for you. From those experiences, you can gather the data you need to make decisions. Once you do decide on a niche, you can identify opportunities you previously didn't even know existed.

FIRST, GET *A LOT* OF EXPERIENCE

As you now know, before you niche up, you should go broad and gain a lot of experience in your industry or market.

You need to have these experiences, to actually try these areas you're interested in and see how they feel to you. Do you feel a sense of satisfaction? Is it personally gratifying for you? Do you even know what you're doing? And are you good at it?

If I've never tried to rehab a house, I may underestimate how difficult it is, how much time it takes, or how much money everything costs. Similarly, as much as someone may think they want to be a professional football player, do they have the natural abilities necessary to excel in that area? Is the area difficult to get into, or is the barrier to entry reasonable?

Additionally, having those experiences creates self-awareness. Let's say I'm watching baseball on TV, and I think, "Hmm, that looks easy. I bet I could do that!" But then I actually play a game and stand at home plate, trying to swing at a ninety-mile-per-hour fastball, like a toddler at his first T-ball game. Suddenly I see just how difficult the game is. Say you're watching NASCAR and think, "That may not be easy, but it's definitely doable. It's just driving in circles; they just have to know how to turn left!" But once you zip into that flame-retardant jumpsuit and put on your helmet, it probably feels a lot more complicated (not to mention dangerous). I'm not saying you're going to hop on the field at an MLB game or get behind the wheel of a race car; these are just examples to show that sometimes you have to have the experience to understand what it's really like.

Okay, now let's look at a more relatable example—let me tell you how this went for me.

Before starting my agency, I performed consulting for a lot of niches—every type of industry you can imagine, in fact. I experimented and tried a lot of areas to discover which worked best for me. I worked with attorneys, plumbers, an HVAC company, e-commerce, even *equestrian* marketing (though, at the time, I'd never even ridden a horse!). I worked with all these areas and saw that they had a real need, that I enjoyed working in them, and that there was an opportunity to do business.

After having all these great experiences, I then decided I should go all-in on one niche…but which one?

Everyone knows that Google helps most businesses, but I could see that it really helps the legal industry in particular. I was reading articles about more and more attorneys graduating every year—attorneys who would need jobs, thus creating competition. With competition, I knew there would also be even *more* opportunity.

However, I—like many people—used to be super intimidated by lawyers. I thought they were more educated than me and, quite frankly, would be mean. Then I worked at a digital marketing agency in the legal space and had positive experiences with most of the attorneys I met. I found that, overall, they were really nice, and I enjoyed working with them. Without having worked at the digital marketing agency, I would have written off working with this entire group of individuals. Never in a million years would I have thought, "Hey, I'm going to help attorneys."

So I launched attorney-rankings.org, the original name of my business, and offered all of the digital services attorneys need—website design, SEO, social media, and pay-per-click (PPC).

After niching to legal SEO, I was sitting in a conference room in Vistage (a business mentoring group) for my business review. Everyone was very complimentary about the company as they went around giving me advice. A few members suggested, "You need to do this for physicians, home services, and chiropractors."

At first I thought, "They're right!"

So because I didn't understand how much of a benefit the niche was, I expanded into these other areas, which slowed my trajectory. I underestimated what I gained from focusing solely on one industry: learning the nuances, copywriting and answering consumer intent, and most importantly, providing value.

Armed with that new awareness, I returned to the legal vertical pretty quickly. Not long after that, I niched up further into the PI space. Once I appreciated the benefits of niching, it allowed me to understand who I enjoyed working with and who I could help the most. In my opinion, personal injury attorneys get a bad rap, mostly stemming from how they're portrayed on TV. On the whole, I've found that they're in the business to help people, not chase ambulances. You're gonna find trolls and negative individuals in any space; in fact, I've been referred to as an ambulance-chaser *chaser*!

To be fair, I had some of these misconceptions about the industry and the people in it too...until I actually worked with them, had my own experiences, and gained awareness. From there, I learned how to help my specific audience (PI attorneys), gaining a deeper understanding of them on a higher level and allowing me to tailor my specific SEO service directly to their needs.

Embracing a niche involves "choosing a side." You're going to be right for some individuals but wrong for many others...and you have to be okay with that.

Before committing to your niche, get your feet wet. Experiment a little bit to see if it's going to work well for you. Instead of just *thinking* you may want to try something, immerse yourself in the situations you're considering dedicating yourself to.

THEN USE THE DATA FROM THAT EXPERIENCE TO MAKE DECISIONS

After starting my agency, I was listening to a podcast, and Seth Godin was on the episode talking about his book *Purple Cow*. In that book, Godin discusses serving the smallest viable market and becoming remarkable *for them*. Most people think *Purple Cow* is about niching, but it's about even more than that—it asks, "How do you become remarkable?" Well, with focus…and oftentimes, it takes niching to accomplish that focus.

That podcast episode triggered me to analyze our client concentration and see who we provided the most value to. I discovered that 70 percent of our revenue came from less than 40 percent of our clientele, which were those in personal injury law. I had suspected that PI was that niche, but I needed both the experience and the data to actually make a decision.

Having these broader experiences gives you more data and information to make decisions about where to niche. Take a look at your data to determine which of your clients are the most profitable—and who you most enjoy working with. Then use that data to help you make decisions instead of just guessing.

We found a natural correlation: we could gauge the value that we provided through the growth of client firms. When we saw that 70 percent of our revenue growth was coming from less than 40 percent of our clientele, it was a no-brainer to niche up into SEO for personal injury attorneys.

> When I started in legal, I didn't know that there was a sub-niche for personal injury attorneys that was much more lucrative until I actually started to narrow down. Now that I'm in the personal injury space, I know there's a sub-niche of PI for mass torts. You can continue to go more narrow into sub-niches, but you won't see these opportunities as clearly until you actually have these experiences.

Once I found my niche, I felt more confident. There's that old saying, "competency builds confidence," and I definitely found that to be true. I could speak confidently to a prospect because I knew them. I'd had experiences that taught me about who they were and what they needed. That confidence directly translated into better sales conversions and more referral work…and I just liked what I was doing better! (We'll take a closer look at these other benefits I discovered as a result of niching up in the chapters to come.)

Competency certainly builds confidence, but it also creates passion. If you're playing basketball and you're just terrible and getting thrown around on the floor, you might not love it. But if you are good at basketball—raining three pointers with the crowd cheering for you—you probably like it a lot more.

I got more enjoyment out of my work in this niche. It was more fulfilling because I could create value in a way that most people couldn't. When you're selling a service, you want to do right by your client. You don't want to steal from them or do a bad job. It's satisfying to know that you can serve and provide value (of course, you want to *keep* providing that service—people aren't going to keep paying you if you do a bad job). I found that the work I was doing in my niche landed exactly in the center of my Venn diagram of purpose, passion, and profit.

> To learn more about this Venn diagram, see "The Hedgehog Concept" in the book *Good to Great* by Jim Collins.

I finally felt like I had a path. When you're marketing to everyone, you're just kind of flinging paint at the wall. You don't really know where you're going to go, or how to improve your business. Once I figured out my exact niche, I could focus on helping these exact individuals. That created a natural trajectory—like the original timeline on the Marvel show *Loki*, rather than one of the variant branches.

To use your experiences to make a decision about your niche, look at the data:

- Who makes up the majority of your clients?
- Who is most profitable for you?
- Whom do you most enjoy working with?

You may see that the percentages aren't there—that most aren't going to buy your services, or there's not a big enough market. When you look at data, you also have to consider:

- Are there enough buyers?
- Are those buyers willing to buy the product or service you provide?
- What competition is out there? How much is it going to take to stand out?

Knowing what I do now, today I would take a different approach to look for a new niche: I would go to census.gov and research how many businesses already exist in that niche, to evaluate how much competition there is and see how much money I could potentially make.

For example, I was thinking about another niche: funeral home marketing. We all kick the bucket at some point, people are not going to stop dying, and there's an immense amount of wealth in funeral homes. They have recurring expenses, so it's not all project work. It's security. It balances.

Then I went to census.gov to see how many funeral homes there are. Who's out there in terms of competition? How much revenue do they make? Are funeral home owners even going to buy digital marketing?

I discovered that this is a really underserved market. It's not a strong niche because people don't like talking about death; it's seen as taboo. So while there are tons of people wanting to open legal agencies, the competition for funeral home marketing is extremely weak—almost nonexistent.

At the end of the day, however, my chief of staff and I decided not to go into this industry because dead dudes in a coffin creep me out and we like the status associated with working in the legal space. No offense to anyone working in the funeral field, but we didn't have the necessary passion for that particular niche.

Once you have the experiences to know where your interest might lie, you can utilize public data to help make the decision.

(If you haven't had the experiences yet, you can still look at the data, of course, but you might not know what to do with it or where to go next.) This is why pilot programs and beta testing exist—to understand if there's a need or demand, and to get feedback on what can be improved.

To be successful in your niche, you need experience, passion, profit, and data. Ask yourself:

- Is there a market?
- Do I enjoy working in it?
- Do I know how to work in it? (Or, am I good at it?)
- Can I get paid to do it?

At the end of the day, you're not doing this just because you're good at it; you also need to make money.

ONCE YOU NICHE, YOU CAN IDENTIFY ADDITIONAL OPPORTUNITIES YOU DIDN'T KNOW EXISTED

Once you decide on a niche, you will then be able to identify opportunities you didn't even know existed before you made that decision.

In 2018, Naval Ravikant, tech investor and co-founder of AngelList, posted a tweetstorm about how to get rich that quickly went viral. In subsequent interviews about the ideas he posted on Twitter, Ravikant made the point that, by specializing, you position yourself to take advantage of opportunities that other people wouldn't have, because of your expertise.

He gives this example:

Let's say you're the best person in the world at deep-sea diving. You're known to take on deep-sea dives nobody else will even dare to attempt. By sheer luck, somebody finds a sunken treasure ship off the coast they can't get to. Well, their luck just became your luck because they're going to come to you to get to the treasure, and you're going to get paid for it.[3]

If you didn't specialize, you wouldn't have the opportunity to dive for that treasure...but because you have taken the time to become an expert—by niching—you get drawn into these unique opportunities.

When I started in the legal space, I didn't realize that PI was such a great sub-niche; I wasn't yet aware of the opportunities the industry offers. If you are starting with home services, you may not realize that working with HVAC and plumbing offers a better opportunity for you than working with electricians (that example is completely hypothetical; I don't know about home services or the opportunities in any of those niches or sub-niches).

Additionally, once I knew my path, I could discover who else was on that same path. I think of it like that scene in *Forrest Gump* where Forrest throws on his hat and just starts running. People follow him because he's going somewhere, and they don't know where to go so his destination is as good as any.

When you're serving one client in your niche, they almost surely have friends and coworkers in the same industry. They're on the journey with you. Your staff is on that journey, too. You're all running in the same direction—and that's what creates

---

[3] You can find more of Naval Ravikant's thoughts on wealth and luck, as well as a link to the original tweet thread, in this collection of interviews on his website: https://nav.al/rich.

momentum. One plus one isn't two anymore; it's three or four because of the way it compounds. Once you find the area that you're passionate about, that you're confident working in, that attracts people who want to work with other people who are good at what they do and care about doing it, who can bring other people in—that's the Forrest Gump effect. He was good at running, and he cared about doing it. He wasn't out there preaching it; he was just running. That's why people joined in.

> In Jim Collins's book *Good to Great*, he describes that journey as being on a bus. You want to get people who don't share your vision *off* the bus and people who do share your vision *on* the bus.

Of course, as we discussed in Chapter 1, by creating that path, you're also blazing a trail for competition to follow. But because they're behind you, they can also push you forward. Think about *The Lord of the Rings*—Frodo has this ring, but he doesn't know anything about it or what to do with it; he doesn't even know where to go with it. As he has experiences, he learns he has to go to Mordor to melt that sucker. People follow him because he has this destination in mind, but it also pulls enemies in behind him—the orcs and Nazgul chasing him make him run faster.

Also, because I was the first one in my niche, I learned things that people who came after me didn't learn in quite the same way, so they didn't have the same awareness. Even though they (obviously) picked a good niche, they didn't have the experiences I did to back them up. They didn't have the data behind their decisions. They could see the profit potential and the opening, but they were not bringing everything we're talking about in this chapter to it.

It looks easy from the outside—they could see that I stepped into this niche and that things were going well for me—but they don't see the work that came behind that.

Sports analogy incoming!

If you've spent the time learning how to shoot three-pointers, you might miss at first, but you'll eventually make the shot. Then, as you keep working at it, you'll make it even more often. Someone watching you, however—even if they stand in the exact same spot and try to emulate exactly what you're doing—isn't going to make those same three-point shots, because they don't have the learning experience behind them.

The people who follow you in your niche may also fall victim to the Dunning-Kruger effect, which is a cognitive bias where people with limited knowledge greatly overestimate their competence. (Remember earlier in the chapter when we talked about how easy it might look to play football or drive a racecar?) When I was a kid playing poker, I thought I was the best poker player ever—until I actually played with a good poker player. (But more about that story in Chapter 3!)

This also happens on the client side. We'll do a really good job on their SEO, and occasionally a client will underestimate our abilities and assume they can do it too. They think, "I know SEO. I've seen Chris do it. I've been in every meeting and heard these reports."

But they're overestimating their abilities because they don't have the actual knowledge. They're not aware that they're not actually good at SEO just because they've seen someone else do it. They'll then either try to do it in-house or get someone cheaper to do it for them—and that dramatically impacts their ability to generate the same results.

Also, if you try to force yourself to do something you're not good at, it's going to be a drain. You're going to spend time and other resources trying to do a passable job at something you don't care that much about—when there is probably something you're already better at, which won't be that much of a struggle.

BUT WAIT, WASN'T THERE A SECOND QUESTION TO ANSWER?

At the beginning of the chapter, I told you that there are two questions you're probably asking right now (that a ton of other people before you have wondered about too).

The first question is "should I niche?" And the answer is still "it depends."

Have you gone broad first, getting a ton of experience, so you have a greater understanding of your industry and market? Have you then used that experience to examine the data, to see what niche may be profitable and enjoyable for you?

If so, then the answer is yes! But the second question is still on the table: "when do I niche?"

The only answer I can give you is that you should choose a niche only after you've had those experiences, *and* you understand what you have a passion for and the competence to do, *and* the data supports you choosing that niche. That's when you will begin to discover those other opportunities that weren't even available to you *until* you find your niche.

Once you have decided to niche up (if that's what's right for you), you'll begin to experience the next advantage: becoming an expert in your niche. Chapter 3 shows you the benefits of that expertise.

CHAPTER 3
↓
EXPERTISE

During the collectible card game boom of the late 1990s, a somewhat niche game called Legend of the Five Rings (or L5R for short) was released to no small amount of critical acclaim. L5R was similar to Magic: The Gathering, but it was unique in the way the players themselves determined the ongoing story of the game's world, through the results of specialized tournaments.

Although I didn't know the term for it at the time, I was successful at playing L5R because I was niching.

The game's setting is a fantasy samurai world loosely based on feudal Japan. As a player, you take the role of the head of one of about a dozen samurai clans, each with their advantages and disadvantages and each with their own colorful name. For example, the Scorpion Clan specializes in trickery and deceit, while the Unicorn Clan is the only one with a significant portion of their military on horseback.

Many players in the game approach it from a generalist perspective. Sure, they may have a favorite clan (or clans), but they tend to sort of...play them all. After all, collectible card games are about continually adding new cards to the available pool, so the "best" clan may change from month to month as new sets are released. And because winning is fun, people obviously like to play with whatever is considered best at the time.

That was not my approach. I chose to play with just one clan: the Dragon Clan. By focusing on just that one area, I could learn all those little details that other people wouldn't notice if they played many clans. After playing thousands of hands, I memorized every single card in my clan. There's no way I could have done that with multiple clans—we're talking about thousands of cards with hundreds of thousands of combinations. I'm good at math, but it would have diluted my focus too much to even try to understand all those combinations of cards.

I ended up being a top-ranked world player and won two state championships in two years. I even got a reputation; I was a known quantity within this world. When people drew my name as their competitor, they would groan, "Oh, man!"

By focusing on mainly playing the Dragon Clan, I put in my ten thousand hours.

YES, TEN THOUSAND HOURS PLAYING WITH DRAGONS

In his book *Outliers: The Story of Success*, Malcolm Gladwell posits the "10,000 Hour Rule." This is the notion that it takes roughly ten thousand hours of practice and repetition to become an expert in any single discipline.

While I probably didn't spend literally ten thousand hours playing L5R (though I spent a *ton* of time playing this game), the time I spent and the focus I gave to this single area allowed me to become an *expert*.

Expertise is important when niching up because people make buying decisions based upon two things: trust and their expectation that the individual they're buying from will achieve the outcome they're looking for. Experts simply have a greater likelihood of achieving that outcome. Look at Michael Jordan—the reason he's held to such a high standard is because if you have him on your team, you're more likely to win that game. Similarly, when a PI firm hires our agency, they are more likely to rank because we're the experts in the field.

Why is this important? Well, when you are an expert in a niche, you gain two of the primary motivators we, as humans, all share: wealth and status.

Naval Ravikant, the entrepreneur and investor who co-founded AngelList, says that wealth is *not* a zero-sum game—anyone and everyone can be wealthy—whereas status *is* a zero-sum game. There's one winner, and everyone else loses. Think about politics: there's a winner and a loser. Same with sports, particularly individual sports. Trials. Even card games like L5R, for that matter. Status is pushed on us heavily in American society, in everything from your chosen industry to the clothes you wear and the car you drive; even the phone you use is a status symbol.

Once you understand that everyone wants both wealth and status, you can see how becoming an expert delivers both. You are the best at what you do, so you can charge more. (This also is the topic of Chapter 4.) And, you are *the best*—the winner—and people want to work with and buy from the best.

> If you think about it—and if you're honest with yourself—you can likely acknowledge that you picked up this book because you are interested in niching, probably for one or both of these reasons. It's okay to admit you want wealth and status. We all do!

Niching up also brings status because you are going against the norm. Most people think you should go for the largest market possible, to make the easiest sales, instead of intentionally limiting your customer base. Doing so actually elevates your status, however, because now you seem exclusive. Seth Godin suggests thinking of it as having a lock. If a ton of keys can open that lock, the keys are not very valuable. But if only one key can open that lock, how much would that key be worth? (Answer: as much as the individual with the lock is willing to pay.) Some experts elevate themselves to a point where they are the *only ones* who can achieve the desired outcome in their niche, and—to connect both status and wealth—if you want the outcome badly enough, you're willing to pay for it.

Additionally, it is difficult to niche up and become an expert so you stand out because you're making sacrifices most other people aren't willing to make. Even when you have natural abilities in your niche, it's not easy. You have to practice, put in the hours and the effort, and put yourself out there. It's certainly not fast; no one becomes an expert overnight.

If you have heart problems and need emergency heart surgery, you're probably not going to ask the heart surgeon how much it costs or shop around too much. You're going to pay whatever it takes to keep you alive because that's the most important outcome. On the other hand, if you have the time and opportunity to choose your heart surgeon, you'll probably ask around to figure out who is the *best* heart surgeon available to you. Who has been operating the longest? Who has done this exact surgery the most times, with the greatest record of success? You don't just want anyone; you want the expert.

PERCEIVED VERSUS ACTUAL EXPERTISE

Niching brings two types of expertise: perceived expertise and actual expertise.

Perceived expertise comes when someone looks at you and assumes that if you choose to do only one thing, you're probably good at that one thing (because why would you choose to do something you're bad at?). If you're going to hire that heart surgeon, they're a specialized doctor—they've put in their time getting their medical training to achieve that status—so there's a perception that they're a great heart surgeon, even if you don't really have solid evidence that they're actually any good at heart surgery at all.

Actual expertise comes when you actually become an expert. You dedicate your time and effort to learning and becoming the best in one area. You give it all your focus and go the extra mile to stand out from the crowd.

Going from being a perceived expert to an actual expert requires focus. When you initially delve into an industry, you learn a lot of things. But to become an expert, you recognize that you can't get to that level in all of those things so you focus—or niche up—into one, specific, smaller area, which gives you the potential to become the best in that area.

One of the most powerful elements to niching is the focus and extreme attention to be the best at this one thing. When Steve Jobs left Apple and then came back, the first thing he did was eliminate a ton of products, to narrow focus on what was most important. However, if he had, hypothetically, decided to expand in several areas at once, creating three different kinds of computers and four different phones at the same time, we might never have seen the iPhone. But he didn't, and because he decided Apple was going to focus 100 percent on one thing at a time, we now have the iPhone that we all know and love.

That was obviously a great decision, but by choosing that focus, they let go of those three computers and three other phones. Choosing to become an expert in your niche comes with its own risks or downsides, and the biggest of those is the opportunity cost to do other things. You spend so much time becoming the best—like Jordan, the best basketball player, or Tom Brady, the best quarterback—time in the gym, time practicing, time playing and establishing that you're the best, and even time spent promoting yourself. But you have to consider if you want to give this time, attention, and focus to this one area—because, if you do, that's not time you can spend anywhere else.

> Billionaire investor Warren Buffet says, "Diversification as practiced generally makes very little sense for anyone that knows what they're doing...it is a protection against ignorance." Most of us have this misconception that diversification is a good thing—but if you *knew* that one stock was better than the other, you would absolutely go all-in on that one stock. The good news is that what you choose to do is something that, unlike the stock market, you *can* control!

I definitely have opportunity costs when I'm away from my wife and my kid, but at the same time, life is about being happy...and doing this work makes me happy. It also gives me the opportunity to do more for my family, to give them more opportunities, and that makes me happy too.

TO BECOME AN EXPERT, LEARN FROM EXPERTS

To become the best, you have to learn from the best. To illustrate this point, let me tell you a story about another card game I learned to play—nope, not a nerdy one (well, not *as* nerdy, at least) this time—poker.

I met Ryan Carter, another top L5R (Legend of the Five Rings) player, at tournaments, and I learned that he was doing internet marketing. Eventually, I moved to Florida to share an apartment with him and some other friends and learn internet marketing.

Well, as these things go, I stayed focused on internet marketing and he went off to play poker.

But Ryan didn't just kinda-sorta play poker for fun; he *played* poker. He got a coach and went professional. He even met his wife on PokerStars (where she had been a pro player as well), and they're still married today.

At one point I told Ryan, "Hey, I want to learn poker, too."

So he became my coach, and we played online poker together, day in and day out, for an entire month. That's all I did. I had extreme focus, and I set everything else aside during that time. In that month, I reviewed every hand I lost—and I played something like forty thousand hands because we were multi-tasking and playing multiple rounds at once (which you can do when you play online, unlike if you play live). We ignored the hands I won and talked through those I didn't. What could I have done differently? Did I actually make the right moves and just get unlucky in this scenario?

> In case you're wondering: of course I even niched here. I didn't learn how to play *all* the different styles of poker (which all have different percentages, positions, and hand ranges). I learned one game: No-Limit Texas Hold'em, and that was it.

Ryan's birthday happened to be at the end of that month. To celebrate, we went out drinking. The next day, we were pretty hungover. Between bites of pizza, Ryan said, "I'd kinda like to play in this millionaire poker tournament on Sunday."

"What's the buy-in?" I asked.

"Two hundred bucks."

"Eh," I said. "I don't know if I've got it in me today."

Then Ryan made things interesting. "I'll stake you and pay the $200," he offered. "Just give me half of your winnings."

"Okay, I'm in."

I played that tournament and, using everything I had learned from Ryan and from reviewing all those poker hands over the past month, I came in fourth place and won $234,000!

A lot of people say, "Oh, Chris, you're so lucky." What they don't see when they say that is that I gave myself opportunities to be in that winning spot. I didn't just play a couple hands of poker; I played *forty thousand* hands in a month, which is roughly the same amount that a person who plays live table games would play in their entire lifetime.

I got a lifetime of poker lessons in a month. That wasn't luck. It happened because I was willing to eliminate everything else, to just focus on playing poker—on becoming an expert—and because I was mentored by someone who was already an expert himself.

When you seek to get better at something, find someone who is already an expert in that subject to learn from. That is the quickest way to accelerate your learning. As author John C. Maxwell says in his book *Leadership Gold: Lessons I've Learned from a Lifetime of Leading*, "It's said that a wise person learns from his mistakes. A wiser one learns from others' mistakes. But the wisest person of all learns from others' successes."

If I had spent the time to learn how to play a hand properly, it would have taken me so much longer. But Ryan already knew how to play these hands, and he'd already put in so much more time—so of course I should follow his instruction.

Everything comes back to time. Why do we give value to experts? Well, as we've established, it's because these individuals have focused on their craft with their limited time on this earth, which is fleeting. They've put in the time and effort. They're not just saying they're the best; they have the experience and results to back it up.

When you are ready to niche up and become an expert, put in the time and learn how to do it. But if you want to be the *best* and level up both your learning and the time that you put in, then learn from people who already have some expertise in that area. Leverage their expertise to gain the only shortcut to becoming an expert yourself.

You can spend your ten thousand hours learning and figuring it out on your own. You can first learn to evaluate what went right and what went wrong, then learn how to do more of one and less of the other. You'll certainly earn your expertise that way. Or you can still have the focus and still put in time, but also have an expert coach you so that you don't have to learn everything from the ground up.

When Ryan was my poker coach, he encouraged me to read certain books and to watch poker on TV, but we also reviewed all of my lost hands. When I played L5R, I was a part of a mastermind group made up of the elite players, so I got to hear what they were talking about. When I started working in affiliate marketing, I was in the affiliate marketing forums, listening to the people who were already the best.

I had to put in that time getting an education to become a specialist, but I also had to leverage the resources that were available to me. Televised poker is available to anybody who wants to watch it, and anybody can find the books I read to learn how to play poker (though not everybody does, and that's why they're not experts!). But then I also had a resource that not everybody has access to, with Ryan as my mentor. That mastermind group and those elite forums also helped me level up because I looked for the resources I could leverage.

Look not only at the resources anyone can access, but also the opportunities that are only available to *you*, whether that's because of who you know or because you're the only person looking in the right place.

EXPERTISE = EDUCATION + APPLICATION

When I got into playing collectible card games, there were two main games at the time: L5R and Magic: The Gathering (MTG). I didn't like MTG as much because players were required to publicly post their decks if they played in tournaments. The way I saw it, I created that deck, and they essentially wanted to take my intellectual property!

I didn't want to publish my deck for exactly the reason we talked about in Chapter 1: when you become an expert in something, and when you find something worthwhile, you naturally bring in competition. I didn't want my deck to be posted because then a whole bunch of other people playing at a tournament could utilize the exact same resource I was, and I'd essentially have to compete against—and beat—myself. All those hours I put into learning this game, and they wanted to just take it and roll with it.

But here's what I didn't understand back then: even if I gave someone my deck, they wouldn't know how to play it, because expertise is about both education and application.

This is the difference between watching a video or reading about how to hit a ninety-mile-per-hour fastball versus going out there with a bat and practicing hitting fastballs. The first piece is education, the second is application—and you need both to become an expert. Watching the video and reading is helpful because it allows you to see the fundamentals, to visualize what it looks like to correctly hit that fastball. Practicing hitting is also helpful because it allows you to physically learn to perform the correct technique. But putting the two together is what will allow you to continue to improve and ultimately elevate your skill level.

You still have to have put in the reps, play the games, and gain the experience. Learn from the best, but don't stop there. *Use what you learn to level up, until you have put in the time and effort to become the best yourself.*

Many people are willing to put in the time and effort to acquire the education and knowledge portion of this equation but stop short of what's necessary to achieve the application side of it.

Remember the movie *The Matrix*? In one scene, Keanu Reeves sits in a chair and has different skills instantly downloaded into his body. When he opens his eyes, he instantly knows kung fu without ever learning or practicing.

Most people who watch that scene think, "Oh, that's so cool. I wish I could plug something into my head and learn a new skill instantly!" But that's not how life works. That plug-in programming chair doesn't exist, obviously.

> At least not until Elon Musk fully figures out his Neuralink brain implants. If he succeeds and you're reading this from the future, how's that neural interface treating you?

You can read all about kung fu, MTG, baseball, or poker, but unless you get out there and practice the moves, the decks, or the hands, you're not going to become an expert.

AS RICKY BOBBY SAYS, "IF YOU AIN'T FIRST, YOU'RE LAST"

From a very young age, my father ingrained in me, "You only play the game to win. Only winning is fun."

My dad was very status-minded when it came to competition (and athletics in particular). As I grew up, I learned to ask myself, "How do I win?"

While I don't necessarily agree with this "second place is the first loser" mentality, I did learn that in order to be the best and win, I had to have complete focus on one thing, whether that was sports or poker or niching. I had to become an expert—and when I did, the status and wealth would follow.

We've already talked about status, so now it's time to focus on wealth. In Chapter 4, we'll look at how becoming an expert means you can charge more—because you're worth it.

CHAPTER 4
↓
PREMIUM PRICING

Let's say, hypothetically, that you have a double chin and you hate it. You need a solution to help you get rid of it. Well, let's take a look at your options.

Option A is to bust your ass, diet, and get a personal trainer to kick the absolute shit out of you every day to lose weight and shrink that pesky double chin. Option B is to opt for a surgical solution, finding a plastic surgeon who specializes in liposuction to suck away the problem area.

(Or option C, as I learned from my first foray into affiliate marketing, is to grow a beard...but not everybody can or wants to rock the face fuzz, so we'll stick with the first two options for this example.)

Which option is easier? That's obvious: option B. In the surgical option, you take a little nap and let an expert do the work. (Obviously it's more complicated than that, and I'm minimizing the risks involved, but you get the point!)

Now, which option costs more? Also option B. Both options rely on an expert—a personal trainer or a plastic surgeon—but the plastic surgeon, in addition to the expertise required to obtain their medical degree, offers something more valuable to *you*, the patient: a faster, easier solution to losing that double chin.

Simply put, plastic surgeons who offer liposuction can charge higher fees because the speed and simplicity of their offer is more compelling.

LIKE L'OREAL SAYS: BECAUSE YOU'RE WORTH IT

In his book *$100M Offers*, entrepreneur Alex Hormozi details what he calls "the value equation." Essentially, the value equation says that value equals the client's dream outcome plus their perception of the likelihood of achieving that outcome (with the least amount of effort and maximum amount of speed).

When you're an expert, your clients are more willing to pay a higher fee because they trust that you have the ability to achieve the outcome they want.

The first reason you can charge higher fees is because your customers are paying for your experience and expertise. You know where to make the incision, hit the hammer, or throw the football. You're not going through the discovery process; you've already learned your skill, and you've already developed your focus.

The second reason you can charge more when you're an expert is *because you're worth it*. You understand the value you're delivering, and your clients will pay more for that value. You are more likely to achieve the outcome they want, faster and with less effort and sacrifice on their part.

> Entrepreneur Gary Vaynerchuk talks about how expertise confers value to things. He uses the example that a basketball in his hands is worth nothing (or close to it). But a basketball in LeBron's hands? Well, that's worth a billion dollars.

Another way to think of this is to consider the Iron Triangle, which says that you can have something that's good, fast, and cheap—but you can only have two of the three at any one time. If you want something cheap and good, it's not going to be fast. If you want something fast and cheap, it's not going to be good. And if you want something that's good and fast, call an expert.

When you're an expert, the "good" side of the Iron Triangle is automatic. People aren't going to pay you for your work if it's not good. Because you are *so* good at what you do, that work is going to be completed faster than it would be by someone who is not as good. That means you have cornered those two sides of the triangle, so your work is not going to be expected to be cheap.

When I began my career in SEO, my retainers were typically in the range of one to three thousand dollars per month. Now, because I have a greater understanding of the competition in my industry and how much I need to invest in order to get the outcome my clients expect, that amount doesn't even cover our minimum fees.

So how did I figure this out?

I used every engagement and interaction that I had with my clients to glean information about the value I was providing through data, feedback loops, and retrospectives. I wanted to know if the value I was creating for the client met or exceeded what I was charging them...and if it did, by how much.

When you work with a variety of industries, however, it's more difficult to understand how much you're worth. I work with personal injury law firms, so I understand why an Uber/Lyft accident would be worth more than your typical, everyday auto collision (because Uber/Lyft typically carry higher insurance limits). Oftentimes, there's more value in a rideshare accident, and I recognize that value because I'm in the industry. I also understand what I'm worth, so I don't devalue my expertise.

All right, so I've convinced you that you are worth the premium pricing you can charge when you niche up and become an expert. Great! But how do you know how much to charge—and when to start charging that amount? We'll focus on that throughout the rest of this chapter.

WHEN TO INCREASE YOUR FEES

In order to increase your pricing, you have to first become an expert so that you're worth it.

But in the beginning, when you are learning your craft and still working to become an expert, charge lower fees and create a low barrier to entry. Once the outcomes that you're striving for start to occur more frequently and more consistently, you'll know that you're ready to raise your rates.

Establish a track record of success and, as you do, raise your prices accordingly.

If you want to become a trial attorney, you can't just read books about being in a trial; you have to get in the ring and try cases. (Expertise = education + application, remember?) When you start consistently winning cases, building a history that you can stand on, you can declare that you are an expert in trials—and charge a higher rate. Your expertise will lend itself to higher contingency results and peer referrals (which is an underappreciated aspect of expertise, which I'll discuss in more depth in a later chapter).

This is another reason why I disagree with those people who recommend niching your business right out of the gate. You have to have all these experiences to figure out what you have a natural competency for, to learn what you're good at, and then to discover how to become consistent at achieving an outcome. You can't just declare yourself an expert and jump into a niche unproven (in fact, the best form of social proof is when *other* people call you an expert, not when *you* do).

Obviously, pricing may vary based on the industry, because the niche you choose dictates how much something is potentially worth to your market. If you are Dunder Mifflin, selling paper to businesses in Scranton, Pennsylvania, they're only going to pay so much. If you're selling *wedding invitation* paper, however (which is still, y'know, just paper), you can charge a much higher fee for the perceived quality of that fancy paper (even if it's just going to end up in the same trash can as the junk mail).

> Chris: "Why is the paper industry so hard?"
>
> Michael Scott: "That's what she said."

As you can see, niching even in something as mundane as the paper industry can affect perceived value.

So once you get to the point where you are ready to niche up and declare yourself an expert, raise your fees to reflect exactly how much you're worth.

COLLECT DATA ON PRICES IN YOUR INDUSTRY

Congratulations, you're ready to raise your prices! But guess what? You still have some homework to do. You have to consider how much your clients or customers will pay to get their desired outcome, which means you're going to have to collect some data.

First, you need data to support your claim of being the best. In sports, you can look at stats to tell you who is the best. In the medical field, you can look at doctors' records to see who is the most successful. That data, in my industry, is all the other personal injury law firms that are ranked number one on the first page of Google. If your clients can look up this information, it is going to factor into how much they're willing to pay.

When you play fantasy football, the really good players go for more money because they make your team more likely to achieve the desired outcome of winning. So when you have two hundred dollars to spend on players, you're going to consider their stats to see who will help you the most.

Once you have the stats to support your expertise, you then have to ask: how much are your buyers willing to pay? And what will they receive for that price?

We talked about status in Chapter 3. Well, it's back in play here because when you are an expert and you charge a higher price, that niche also improves your status *and* the status of your customers. You might buy a cheap watch just to tell time, but if you buy a Rolex, you're more likely to achieve that outcome of status because it's an exclusive, high-priced item made by specialist watch manufacturers.

When considering pricing, you have to figure out the benefit that your customer or client will receive—how it will help them achieve their desired outcome. The benefit of a generic pen is that the user is able to write stuff down. On the other hand, a fancy gold-plated pen that comes in a special holder confers status. In fact, it may never be used to write at all, but instead remain only a display piece.

> When we buy something, we're not buying a feature, we're buying the benefit, the outcome. To explain this, I'll paraphrase an example given in an episode of the *My First Million* podcast, which put it very well:
>
> If Mario (while running through the Mushroom Kingdom on his way to rescue the Princess) picks up a Fire Flower, it turns him into a new form of Mario, one that can throw fireballs from his hands.
>
> If someone asked me to sell them on the Fire Flower, I certainly wouldn't focus on how it can grow in tough environments or how pretty the petals are. I would (obviously) focus on how merely picking it up enables me to *throw fireballs* and nuke those little mushroom bastards.
>
> The benefit is the whole *point*, man.

You can charge more because you're worth it, but you have to believe that you're worth it. You have to put it out there that you are worth it. You have to show your buyers that you are the best—and they'll believe it (because you are). Then, when you back it up and prove that you're the best, with the data showing your expertise and experience making it more likely that your clients will achieve their dream outcome, they are happy to pay more *and* to have *their* status elevated for paying that higher price to have the very best.

When you're having heart surgery, sure, you want the specialist, the cardiac surgeon, but you want the head of cardiac surgery because they have the additional expertise that has allowed them to obtain that status. Not only that, you want the head of cardiac surgery at the *best* hospital—price be damned.

A word of caution: not everybody is going to be willing to pay what you are worth. That doesn't mean you should lower your prices; that means you are niching correctly! The goal is to find the smaller segment of the market that will pay the higher prices because they also understand the value of the service or product you provide.

TEST WHAT THE MARKET WILL SUPPORT

Of course, you'll eventually hit a point where people—even your biggest supporters—won't continue paying more. At that point, you'll fail to convert because (at least for most of us) whatever we do is not something people will pay literally any amount of money for. There's always a point of diminishing returns.

It's at this point that many entrepreneurs make a critical mistake. The solution is *not* to reduce your prices; it's to *increase the value* that you provide.

You have to test what the market will support, and I think you should do just what I did: continue to raise your fees until you hit the ceiling at which the value you provide is not enough to exceed the pricing…and then increase the value to justify the higher fees.

This is all just about perceived value and math, and that math is tied to time. Time is fleeting and, without trying to get too dark, we're all going to die. So the most precious resource we have is time, and that's what we're all selling, ultimately.

Again, what makes a good offer? You're going to achieve the outcome faster. Why will people pay more for it? Because their time is fleeting, too, and they'd rather spend their money to buy your time. You've put in the time to gain the expertise. If somebody else wanted to, they could probably learn SEO or brain surgery, if they were willing to go back to school and put in the hours. But they're not—that's what they're paying you for!

So they're already paying you X amount, because they don't want to put in the time to learn how to do this, and then they'll pay you even *more* because you're the best one doing it. Seriously, go out there and find out *just how much* clients are willing to pay for those hours and *your* expertise.

WHICH COMES FIRST, THE SELLER OR THE STATUS?

You, the seller or service provider, have to have some kind of status so that people will buy your product or pay for your service to get that status for themselves.

In Jordan Belfort's book *Way of the Wolf*, he says that you have to convince your buyer to have trust in three things: you, your company, and your product.

People put their trust in you because of your status. It's those awards on your shelves, the degree on the wall, the plaques on your desk. Why do you have these? And why do people care about them? Because they are symbols of status, and status declares to the world that you are the best. By being the best, you can charge more because you're more likely to achieve the best outcome.

It's a never-ending loop: first, you have to achieve the outcome. Then you have to be the best at achieving the outcome, to get status. And then, because people see that status and you back it up by actually achieving the outcome faster, they'll pay you more. And then by achieving the outcome even *more*, chances are they'll continue paying you *more*.

Does your brain hurt yet? I know mine does.

Let me put it more simply: niching up creates opportunities and can create abundance instead of scarcity. You would think that shrinking your market means you would have less opportunity to make money, but in many cases it actually helps you make *more* money.

Now you know that you can make more money when you niche by charging premium prices. In the next chapter, we're going to look at another benefit of niching up, which also ultimately leads to higher profits: increased conversions.

CHAPTER 5
↓
CONVERSIONS

My family recently moved into a new home in a new neighborhood. The other night, my wife and I were trying to decide what to have for dinner. I took out my phone and hit the "food near me" button.

As I looked through the list of restaurants in our area, I saw *tons* of fast food places: Burger King, Taco Bell, McDonald's. That was to be expected. Then, out of curiosity, I decided to see how many nice steak joints were nearby: one. Also not a surprise, but disappointing when you really want a premium piece of meat!

Think about these two extremes. Which restaurant do you think has better food? The steakhouse. Which one costs more? Yup, again, the steak place. And who has less competition? You guessed it again! It's the specialized, niche restaurant.

The fast food restaurants have a *ton* of competition. That's why they have to sell a million Happy Meals; they're all trying to reach a general audience. When you're selling a five-dollar combo meal, you have to do massive volume; when you're selling a fifty-dollar steak dinner, however, you only have to reach that smaller subset of people who want that higher-end experience.

I'm not saying one of these is better than the other. If you have a hungry kid in the backseat, you're going to hit the drive-through on the way home. That kid doesn't even want a steak in the first place: she wants a Happy Meal!

But when you want a steak, if you're like me you want a *good* one. You're going to ignore all of those burger joints and head straight for that high-end joint because they know what you want and they can deliver that juicy, meaty deliciousness cooked to medium-rare perfection. You get to undergo a conversion—from hungry to happy—and the steakhouse also gets a conversion by taking you from potential consumer to satisfied customer.

YOU'RE NOT A JACK OF ALL TRADES; YOU'RE A MASTER OF *ONE*

In the last chapter, we talked about the ability to charge premium pricing as an expert in your niche. This chapter focuses on another advantage of niching: having a higher rate of conversions. Even at those higher prices, you will be able to convert more of your prospects to customers because you know what you're worth, you know what it takes to achieve the outcome, and you've shown that you can deliver value. Prospects tend to see that and think, "They cost more because they're worth it."

In general, there are three ways to generate more revenue:

- get more leads
- increase your win-rate
- raise your prices

Where niching really has an impact is on your ability to charge higher fees (as we've discussed) and on your win-rate percentage—winning more leads through conversions.

With niching, as you know, there's a cap on your market. But because you become an expert, because you niche, you can have greater conversions and raise your prices. If you're more of a generalist, you may be able to increase the number of leads, but you'll likely have a harder time converting those leads and raising your prices because you're not seen as an expert in anything.

As we've established, every sales conversation is really a conversation about trust. You make a promise about what your product or service can do, and the person you are selling to trusts you, or your product or service, to deliver on that promise (or not). By niching, you are aiming to show that prospect that you have experience, focus, and expertise in your area, so they are more likely to trust you. Because of that increased trust, you are more likely to convert those conversations into sales.

Even more importantly, you don't just offer promises; you have concrete *proof* that you can give clients the outcomes they want because you have a historical track record of helping people like them. We actually use that as a tagline on our website: "Anyone can make you promises...We can give you proof."

Because you are an expert in that one area, you can also relate more to your target audience. In sales conversations, the prospect often asks, "Who else like me have you helped?" When you niche, *all* of your customers are like that prospect to some degree, so you can confidently say, "We've helped a hundred other people dealing with the same problem you are, and *here's how*."

That confidence is acquired by understanding your prospect and knowing exactly how to deliver the results they want (in other words, by niching up). That completes the loop: you know what you are doing, so you have confidence; your prospect sees that confidence and develops trust in you; then, after converting them to a customer, you give them the outcome they want; and finally, your confidence is cemented both in your actual expertise (you did the thing!) and in your customer's perception of your expertise (wow, you did the thing for me, so now I trust you even *more*!).

Here's the secret: once you niche up, these sales conversations won't be nearly that difficult for you because you will be spending all your time working in that one area, with the same customer base. You'll have a fairly precise notion of exactly what a new prospect wants because you're always working with people like them. It's *so* much easier to talk to somebody whose problems you know and understand than it is to sell to somebody who is an unknown quantity because they come from a wider, more generalized customer base.

If I help somebody with SEO in the general legal industry, I only have an *idea* of what they want. But because I specialize in working with PI attorneys, I know *exactly* how to help them because I have legitimately worked with hundreds of other PI attorneys. There may be some small differences (e.g., practice areas, locations) in what they need, but overall, they're doing basically the same thing as everyone else I work with.

> Think about choosing characters to play Dungeons & Dragons. When you see someone who is a Fighter/Wizard, you don't think they're the best fighter *or* the best wizard; you probably think, well, they can do two things, but they're also mediocre at two things. They're dual class but they only get half of the skills from each. On the other hand, when you are the person who is just the wizard or just the fighter for five or six levels—you're likely *way* better at spells or fighting than the person who is only halfway decent at both.

Additionally, niching up allows you to be more specific, which also helps increase conversions. Because you're solving a problem for one smaller group of people, for whom your product or service is designed, you can easily put words to that problem. When a prospect comes to me, I can speak directly to them, as an individual, and target specifically what they're looking for. Rather than saying, "We help businesses rank on the first page of Google," I can say, "We help your personal injury law firm rank on the first page of Google."

Being specific makes you substantially more memorable and creates the perception of deeper expertise. I was on a panel recently, and we were each asked to introduce ourselves with what we do.

The first person said, "Hi, I do SEO for law firms."

The second person said, "I do SEO for family law, criminal defense, and PI."

Then it was my turn. "Hey, I'm Chris, and I do SEO for personal injury firms."

For a *personal injury attorney* listening, there's only one clear choice. I stand out because I can speak directly to those—and only those—individuals. (Let me be clear, the other two agencies are great. My point here is only to show how the way in which you position yourself as an expert in your niche leads to how you're perceived as an authority.)

Finally, when you put all of this together—the trust, the proof, the confidence, and the specificity—niching helps you develop rapport with prospects more quickly, which makes it easier to convert them to clients. If I'm doing a sales call and a personal injury attorney says, "We only do high-end litigation, so we just want serious injuries," and I reply, "Oh, so eighteen-wheelers, brain injuries, birth injuries, wrongful death…" and rattle off a list of what they are looking for, they automatically think, "Wow, he knows his stuff." In fifteen seconds, they can see that I'm an expert in this space and that gives them a pretty good idea that I understand their pain points already.

If you don't have the experience in a niche, most of the time you can't do that. By niching, you show that not only are you a part of their world, you've intentionally *chosen* to *become* part of that world. The prospect sees that you understand their problem and you know what they're going through or looking for, which makes it even easier for them to trust that your solution works. Then you back that up with a proven solution that you're confident in because you've delivered it to so many other people *just like them*.

CREATE CASE STUDIES AND TESTIMONIALS

One easy thing you can do to increase your conversions when you niche is create case studies and testimonials specific to your avatar's pain points. Use evidence that supports you being an expert so prospects can see how you've removed that pain for your existing clients.

Just getting started? Do the work for free if you can. Everyone has to start from somewhere to move up the ladder in regards to bigger engagements.

It's all about trust, trust, trust.

After you've done the work, have a conversation with your client. Use that information as a feedback loop to improve. When you've reached a point where you have provided sufficient value, ask for a testimonial so that client can share their experience.

It's that simple. Testimonials are easy for people to write: they just have to lay out their experience working with you. It's not intrusive and requires very little coordination.

If you want to create an even more powerful example to show prospects, you can ask your clients if they would be willing to give you a video testimonial. In a video, leads can see the individual giving the testimonial—they can see that it's a real person. They may think that anybody could be writing a testimonial, but when they see somebody's face attached to it, they feel like they're hearing the experience of a real human being who's like them, who used your service and got the same results the prospective client wants.

To go a level deeper, you can also create case studies. A case study takes it a step further than a testimonial because it shows your prospects an overview of the whole journey. A case study may encompass video, proof items, process, and deliverables. Basically, a case study shows where someone you helped was *before* you helped them. What problem or pain point did they have? Then it shows where they are now, *after* you've helped them. How did you solve that problem? What were the results? In the next section, we'll talk about using a client's specific language to fill in these details.

The goal here is to show as much evidence as possible that you've helped a lot of other people, just like the people you are selling to. Share awards, testimonials, and case studies specifically from or about other people in the same world as your prospects.

Just a tip? No one wants to be a case study, but everyone wants to be *featured*. So instead of asking, "Can you be my case study?" which sounds like you might dissect them, consider saying, "I would love to feature you and talk about your story."

When you phrase it like that, how could anybody say no?

WRITE COPY USING THEIR LANGUAGE

Another way to increase your conversions is to write copy using the language people in your niche use.

Words matter. When I write or speak to my target audience, I don't use the word "lawyer." I say "attorney" because that's how they refer to themselves when they communicate among peers. Moreover, we refer to them as "the preeminent attorney" in whatever field. The word "preeminent" is not commonly used in most industries—but it is in the legal vertical.

Get to know your audience's pain points and their desired outcomes, then use their specific language to speak to that pain and promise those outcomes in all your materials. I know most PI attorneys don't just want small, fender-bender auto crashes; they want serious injury cases. When you have a deep understanding of your audience (which is easier to do because that audience is smaller when you niche!), it helps you write copy that speaks to that audience.

Similarly, I don't say "grow your business" or "grow your revenue" in my copy, both because those aren't the words they use and because those terms are not specific—*how* are you going to help them grow their revenue? With my audience, I talk about "signing more cases."

If you were going to write copy for a car dealership, you wouldn't say you are going to "generate revenue"; you would say you are going to "*sell more cars.*" Maybe you'd get even more specific and say, "I'm going to help you sell more used Toyotas." If they sell more cars, they're going to make more money; that should be assumed. Show them that you understand *how* they're going to increase their revenue when you say, "My product or service will help you sell more cars."

This also helps differentiate you when everyone else is talking about revenue and growing their business, but you're speaking to them about the outcome they care about. You're using their language—that's going to make them take a closer look at you, because you seem like the better fit.

This may sound nitpicky, and these may seem like little details, but all of these smaller elements have a compounding effect.

BE CONSISTENT

Another aspect of niching that builds trust, and therefore increases conversions, is having continuity and consistency in your offer and communication.

Blair Enns, author of *Pricing Creativity*, asks you to imagine you went to a new barber. The first three times you got a great haircut, but the fourth time it was *terrible*. You probably wouldn't go back to that barber, right? He didn't give you a consistently good haircut.

Now, your prospect doesn't know whether or not they're going to get a good haircut every time. But when they see you talking about haircuts the same way, using words they know, showing that you understand what they want, *and* they can see the testimonials or other results from people you've helped—all of this paints this picture for them. It helps them understand you are their person.

Again, it comes down to trust. You will have more conversions because the individual is more likely to trust that you're going to deliver the outcome that they're striving for, because you've done it many other times and you have experience in their niche.

#WINNING

We track win rate in our CRM (customer relationship management) software, so I have statistical data to show when our win rate increased, and then I can look at what the common denominator was, what we did differently.

We had great results previously, but when we made the change to niche and really showed that from a public perspective, our win rates improved dramatically—and, subsequently, our fees increased. In 2018, we had eighty clients and we made three million dollars. Fast forward just two years later, to 2020, and we had only twenty clients, but we made six million dollars.

Sales calls became even easier when we started converting at the higher rate. They were more consistent. When almost every call is the same, we were better prepared and able to craft a better presentation with a better offer—one that better helps our target audience because we were no longer solving for an average of everyone; we were solving for a specific individual. We knew what their pain point was, so we knew how to create an offer that spoke to that pain point.

As an example, spam is very prevalent in the legal vertical, so we had to hire dedicated employees to fight spam. That made our offer better because we were proactively reporting spam to help our clients move up in search engine results. Spam is not as much of a problem in most other industries, so if we weren't in this niche, we wouldn't have known what a problem it is for our audience, let alone have a plan to address it. But because we know our industry, we know that it's a problem and already have a plan in place, so we can talk about that in sales conversations, leading to more trust and higher conversions.

When you get to know your niche, you can similarly get to

know your customers, plan ahead for problems you know they will have, and build trust even *before* they become customers.

Creating more trust also leads to another benefit: better relationships. Chapter 6 will show how your relationships can become currency.

CHAPTER 6
↓
RELATIONSHIP EQUITY

Have you seen the show *Justified*? It's a Western-style modern crime drama starring Timothy Olyphant as a federal marshal named Raylan Givens. (Mild spoilers ahead!)

There's a scene in the first episode of season four where Givens has captured a fugitive named Jody Adair and is transporting him in the marshal's car while Adair complains nonstop. Givens is on a call when Adair starts up again, saying, "I think my arm's falling asleep."

"Stop talking," Givens says.

Incredulous, Adair responds, "I'm not gonna stop talking!" He continues, "Hey, you gonna have to turn me in unharmed, or you gonna catch some shit."

Givens slams on the brakes, causing Adair to smack his head on the dashboard. Then he says, "You know what your problem is? You got no self-awareness. You think trying to do right by your children excuses everything—even killing men."

Adair replies, "They were heroin dealers. If they'd just given up their money, none of this would have happened."

"Every problem, that's someone else's fault." Then Givers delivers one of the most memorable lines of the series: "You ever hear the saying, 'You run into an asshole in the morning, you ran into an asshole. You run into assholes all day—you're the asshole'?" (And then he shoves Adair in the trunk!)

That famous line can apply to niching as well. If you work with one attorney, you just worked with an attorney. But if you work with attorneys all day, you're just like them—and that's your niche.

To put it another way, hanging out with one attorney is just happenstance, two is a coincidence, but three is a customer base.

EARNING INTEREST

Einstein famously said, "Compound interest is the eighth wonder of the world."

When most people read that quote, they automatically associate it with finances and stocks, but there's a different type of currency that also earns compound interest, and that's relationship currency (also known as relationship equity).

According to an *Inc.* article from 2016, the definition of relationship equity is "the distribution of resources between relational partners." The article goes on to say that equity theory "examines whether what the parties in a relationship get out of a relationship is equal to or greater than what they put into it."[4]

---

4   https://www.inc.com/adam-fridman/three-reasons-relationship-equity-is-the-new-lead.html

You can think of relationship equity just like the equity in a house. When you start a relationship with someone, whether personally or professionally, there is a mortgage on that relationship. You owe a debt of value. So you provide value in some way, which makes a payment on that mortgage. (We'll look at ways you can provide value in your niche a little later in this chapter.) As you continue investing in the relationship, you're paying that debt of value and therefore gaining more and more equity.

An important point to make here is that you're not looking to get something out of that relationship immediately; you're giving first. Relationship equity is not about immediate value, it's about long-term relationships, which then lead to that compounding interest.

> In his book *How to Win Friends and Influence People*, Dale Carnegie says that the most powerful word you can use when you meet an individual is their name. If you're new to an industry, how many people know your name? Not many. But the longer you're in an industry, the better known you become—and the more relationship equity you develop.
>
> To put it another way, Michael Mogill, author of *The Game Changing Attorney* and one of my acquaintances, says, "Best-known is better than best every time—and the best-kept secret is broke." If people don't know you, then they're not going to work with you, no matter how good you are. The goal is never to be the best-kept secret.

When you are developing these relationships in your niche, you may start by working with smaller businesses or clients. As you gain experience and establish a track record of success, people in your market will grow familiar with you, so you can be seen as trusted. When they trust you, they are more likely to introduce you to their peers. This is how you can become a thought leader.

Those relationships stack, so as you develop and work with larger firms, you get even more trust and continue to move up the ladder. Essentially, you're niching up through your relationships and experience.

All of this creates a world in which you are surrounded by the people you want to work with and those people want to work with you because they recognize you as an expert.

But in the beginning you don't have those relationships. You don't know who the power players are. If I were to start in the home services industry, I know a handful of people I can contact to begin building that momentum. But in the PI space, I know *hundreds* of people.

Additionally, you'll find that, in any industry, the higher up you go, the more difficult it is to get in front of people. If I'm working with a solo practitioner or small firm, I'm probably talking to the owner, but as I move up to these larger organizations, I may talk to a marketing advisor first, then a CMO or different partners, before I really get to the main decision-maker. Getting to the top of that ladder requires an immense amount of trust because there are more barriers and more people who have to trust you, so you have to have even more relationships for them to see your authority.

All of these relationships, this trust, and the authority that you build as you develop equity combine to form your reputation in your niche. But not only do you have to build your reputation, you have to *keep* it as well. As Warren Buffet says, "It takes twenty years to build a reputation and five minutes to ruin it. If you think about that, you'll do things differently."

You are spending a lifetime building your reputation and relationships in your niche. You are not getting a reputation in one industry one day and then moving on to another the next day. This is a long-term investment into those relationships upon which your reputation is built and to generate relationship equity.

How can you do that? That's what we'll look at in the rest of this chapter. I'll show you how to first get to know your people by joining them where they are, then to provide value *before* making a withdrawal from your relationship equity, and finally who *not* to associate with (hint: it's the people who want to take without giving).

I WANT TO BE WHERE THE PEOPLE ARE

In order to build relationships, you have to meet people in your niche—and that means going where those people are.

Join communities or associations your target audience are members of. Go to the conferences they go to, and spend time where they congregate. Once you get to know people and do business with them, your relationship network can grow exponentially. Your clients can then introduce you to their associates at those conferences.

> When you're in an industry, it becomes like the Six Degrees of Kevin Bacon. *Someone* knows how to connect you to the person you want to talk to; it's just a matter of finding the right people to make those connections. Once you do, you'll find that the number of people separating you from your target audience gets smaller. If I wanted to talk to John Morgan (who owns the largest personal injury law firm in the United States), I know exactly who to ask for an introduction because I've already gone up the ladder to get to know the person just below him. I wouldn't have to go through six people; I could just go to one.

This again shows the compounding effect of relationships. If you know five people (who you have helped and given value to first, as we'll see in the next section), and they each introduce you to someone at a conference, you now know ten people. If, at the next conference, those ten each introduce you to two people, you've gone from five people in your circle to thirty in the span of two conferences. The more people you know, the more opportunities you have for introductions or referrals (the topic of our next chapter!).

"But Chris," you may be thinking, "how do I know where to find them in the first place?"

Excellent question. When you are first building these relationships, you likely won't know where your prospective clients congregate or what conferences they even attend. So how can you find your people? By using your niching superpower: your focus.

Focus on finding out everything you can about the people in your niche's market. Follow them online, add them to your social media networks so you can see what they're interested in. You'll also learn where they're congregating, which associations they're part of, and what conferences or events are coming up. Discover where they're engaged, what their pain points are or needs they have, and what solutions they may be looking for.

If you see a bunch of people you follow posting pictures on Instagram of this cool conference they're all at, make a note so you can plan to be there next year. By then, you can have even more relationships established so that when you go, your people will be able to introduce you to their networks.

BUT DON'T BE CREEPY

In the next two sections, we're going to look at more specifics about giving value before you ask for anything and how not to be someone who takes without giving, but I want to highlight the importance of both of those here, while we're talking about meeting new groups of people, whether in person or online.

You want to be the person who adds value to a new relationship, the expert in your niche; you *don't* want to be the creepy stalker who takes without giving value. They follow people because they get something from watching others, whether that's information or status by association. (This is why outbound marketing is so challenging—some creep you don't have a relationship with keeps emailing you to try to get you to buy their product.)

So be a fan, not a stalker. A fan gives praise, accolades, and validation. If you start following someone in your niche on Facebook (or Instagram or LinkedIn or wherever your people hang out online), don't be that person who just looks at their pictures all day. Comment on their posts, engage with them, and try to be helpful. If you're in person at a conference, go up to them and have a conversation about a recent win or something positive about them.

But if you want to be Creepy McCreeperton, ignore this advice.

DEPOSIT MORE THAN YOU WITHDRAW

Just like you want to deposit more into your bank account than you withdraw, you don't want to overdraft your relationships either. That means giving more than you take.

You wouldn't walk up to someone you've just been introduced to and say, "Buy my services from me today!" before you've even said hello, right? (You shouldn't do it right after you say hello either, to be clear.) You also wouldn't approach someone and ask them to marry you right after you meet them. You have to court them first. You have to build trust. You don't lead with an ask. You get to know them, develop a relationship, and provide value.

That takes time. You've probably noticed that we've talked about time in nearly every chapter, and that's because time has value. Relationship equity means taking time to meet and get to know all these individuals in your niche. It means showing them, time after time, that you are a good person to know because you can help them—without expecting anything in return (except, wait for it, *their* time).

Only then, after you've helped an individual several times, should you consider making a pitch to them.

> In Gary Vaynerchuk's book, *Jab, Jab, Jab, Right Hook*, he describes jabs as the things you do to provide value first so that when you make an ask it's not so weird or intrusive. (He also specifies that you've only earned the right to *ask*—that doesn't automatically mean they have to say yes.)

When you lead with value first, it lowers people's guard. It shows that *you* are a person of value, not just a taker. That can be an opening to a relationship. If you give value, other people are more likely to value you.

I have a podcast, called *Personal Injury Mastermind*,[5] because a podcast is a good way to give value to a lot of people at once. One episode can reach thousands of people, which pays that debt of providing value and therefore gaining equity faster and more efficiently than talking to each of those people individually. The trade-off is that you earn interest at a lower rate. When you build personal, one-on-one relationships with people, you earn relationship equity at far higher rates than you do through one-to-many channels.

When I ask someone to come on my podcast, I'm not asking for their business; I'm asking for them to be a featured guest on my show. That is a value-add because we talk all about them. At the end of the show, I don't ask them to sign up to be an SEO client; I promote their episode. I promote *them*.

Later though, when they're considering who to work with for SEO, I hope that they'll remember me and think, "Chris was nice. He wasn't a sleazy salesperson, and he didn't try to hard

---

5   At pimm.fm, if you want to listen.

sell me. I liked the conversation I had with him, and I'd like to work with him."

WRITE CONTENT

One of the best ways to gain a reputation for being someone who gives value is to write content specifically geared toward the needs of the people in your niche, showing that you really understand them.

Content can mean a variety of things: social media posts, blogs, videos, podcasts—whatever media your niche is most interested in and most likely to consume. Podcasting has been one of the greatest outlets for me to build relationships because I'm giving value to my guests by giving them a stage without asking for anything in return.

In Chapter 5, we talked about writing marketing copy targeted for your niche. Here, you use that same knowledge to write different kinds of content that also provide value. All of it is showing that you know them. You know what their problems are, and you know what would give value to their businesses and their lives.

As the people in your industry consume your content, it is likely that they will come to value you. They'll learn who you are and that you just give away valuable information. That builds trust, which you've seen leads to more conversions (and, as you'll see in the next chapter, also leads to more referrals).

One of the first things I did to start building relationship equity in my niche was to create value on Google+. (In case you don't remember Google+, it was a relatively short-lived social media network where you could create and share lists of contacts, which they called "circles.") I started a law firm networking circle where I curated the list and shared it with anyone who joined and participated.

Even though I wasn't providing a service or selling anything, that list became immensely valuable because I served as a connector for people in the circle. But that list was only valuable because I spent time engaging with all those people on social media, getting to know them, and having them introduce me to more people in the industry.

I built my podcast the same way—meeting a few people who were connected to and trusted by my niche, and building relationships slowly.

Sadly, I got ghosted for my third podcast episode. I had researched the guest and scheduled time with them, and they just didn't show up! However, that's the only time that's happened, because I've since done more than a hundred episodes so now I'm more trusted. My podcast is more legitimate. In fact, while I had to email people to ask them if they would be on my podcast for the first fifty or so episodes, now people email *me* to ask if they can be a guest.

When you develop relationship equity, people will want to interact with you. As you build that reputation of being a person of value, you're no longer just somebody looking to meet other people. Other people get to know your name and want to be introduced to *you*.

DON'T BE A SOUL-SUCKING SNAKE OIL SALESMAN

In the SEO space particularly, certain people are referred to as snake oil salesmen. You don't know them. They travel from one industry to the next, selling their wares. It turns out their solution doesn't actually work—but they don't care because they've already moved on. They're gone, and so is your money.

If you choose to niche, you're choosing relationships. You're choosing to be loyal, to build a strong, positive reputation. That's not to say there are no snake oil salesmen in niching, but they are few and far between because they can't build their reputation. No one trusts them, so it just doesn't work. Niches are small groups of people, and they're going to talk to each other. If someone has a bad reputation in a niche, no one's going to work with them. If you're great, people are going to talk about you—but if you're awful, they're going to talk even more.

I've had to part ways with more than one client because of reputation-related problems that arose while working with them. If we had kept them on as clients, it would have affected our other relationships because we'd be seen as guilty by association for continuing to support them.

Your good name is connected to everyone you work with. Once a client allows their reputation to take that hit, we have a choice: cut them loose, or allow them to drag us down with them.

There are larger, more negative repercussions for being associated with them, so it just isn't worth it because niching isn't only about money. It's about everything we've talked about: relationships, reputation, and value. We never want people to look at us and think, "They do SEO for *that client*. They don't have any integrity about who they choose to work with; that's not the right person for me."

We would run the risk of missing out on working with future clients because they want someone whose values better align with theirs.

As motivational speaker Jim Rohn says, "You are the average of the five people you spend the most time with." Take a look at your niche: you are the average of your five most well-known clients.

Just as you don't want any of your clients to drag down your average, I have to come back to the point that you don't want to lower *their* average either. This chapter is all about how to add, to give, not take away. People don't want to have a relationship with those who just take from them without giving anything in return. (That's why we talked about giving, giving, giving before you even think about asking for something.)

You've probably had the experience, when going to conferences or events, of seeing *that one person* coming in your direction and doing everything you can to avoid them. They feel like a dementor from *Harry Potter*, someone who will suck your soul and drain all your energy—in this case, by wasting your time, leeching any information they can get from you, and sapping all happiness from the room, leaving only dark clouds and coldness. (They definitely didn't get the memo about not being Creepy McCreeperton).

Ask yourself, if you suddenly found yourself transformed into a *Harry Potter* character, who would you want to be: the dementor who sucks the energy and souls from the people around them, or someone like Fred and George Weasley, who give away candy and magic tricks? (Sure, they're turning their audience into magical test subjects, but hey, free candy!)

Or do you want to be someone like Hagrid, who tries to elevate and boost every person (and even animal) he comes into contact with, who teaches people about misunderstood magical creatures, and who will show up with a cake on your birthday to take you away from the horrible family that has been abusing you ever since your parents were killed while trying to save the world from evil?

I may have gotten a bit off-track with that metaphor there, but honestly ask yourself who you want to be. Say you're just starting out, choosing a niche, making a name for yourself, and building your reputation. Once you niche up, develop relationships, and establish trust, who do you want to have around you then? Don't be a dementor, and don't allow yourself to be associated with them either.

RISING TIDES LIFT ALL BOATS

In the previous chapter, we talked about how every sales conversation is based upon trust. When you are better known in your industry, when you give value and build relationships, you are more trustworthy than someone who is an unknown entity. You have a higher chance of converting your prospects and of being introduced to other people in your field.

Well, a rising tide lifts all boats, and developing that relationship equity also lends itself to giving and getting more referrals, which is the topic of the next chapter. When you know the industry, and you know who the people in your niche should talk to, to solve their other pain points, you become seen as even more of a trusted advisor.

CHAPTER 7
↓
REFERRALS

It feels good to help people.

The other day I was pressed for time and had to quickly run into Dollar General to get a birthday card. I got in line behind four kids who were buying chips, soda, and candy. As the cashier rang them up and gave them their total, the first kid paused, dug through their pockets, and said, "Oh no!"

The other kids, who were a little younger than the first, asked what was happening.

"I lost the twenty-dollar bill," they replied.

You could see they were devastated. That twenty dollars represented an absolute haul of snacks for them, and now they were facing a future of scarcity.

Before I could step up and pay for their food, they said, "Be right back," and ran outside to look around the grass out front.

I paid for my card and walked out. Two of the kids were looking in the bushes while the other two were on their hands and knees in the grass, combing every inch of the area for their missing money. All four of them were visibly upset.

I walked up to the oldest kid and asked, "Did you find that twenty?"

"No," they replied sadly.

"Here," I said, and I handed them a twenty-dollar bill from my wallet.

"Oh my God, *thank* you!"

The kid ran over to show his friends, and they were all so *happy*. Twenty dollars is a lot of money to a kid; it represents so many possibilities. (I'm not saying twenty dollars is nothing when you get older, but losing a twenty-dollar bill was a huge deal for those kids at that time, and I was fortunate to be in the position to give them the money.)

And I was happy to help them because I want to live in the kind of world where a person tries to make somebody else's day better instead of worse.

WHO YOU GONNA CALL?

In the previous chapter, I asked you to decide whether you wanted to be a dementor or Hagrid, a taker or a giver.

Consider another question here: do you want to live and work in a world where people are just looking out for themselves, where the client wants to get the most work for the least money, and the person helping them only does the bare minimum and never sends a client to another professional that may be better suited to help them? Or would you rather be in a niche where everyone is trying to solve problems together and looking to help each other, whether that means doing something yourself or connecting someone with the right person to achieve their desired outcome?

I know which I prefer, and it's not even close. Helping people makes me feel good, and I think it just feels better to have positive relationships with the people around me.

A great way to help people and add value (which, as we also saw in the last chapter, builds relationship equity) is to give referrals. You won't be able to solve every problem your target audience has—especially because by niching up you're choosing to focus on becoming an expert in one area instead of just doing an okay job in several areas—but when you can connect them with someone else who *can* solve that problem, guess what? The outcome is the same: the problem gets solved (bonus: now their trust in you likely grows, which helps increase retention).

The best way to create a referral relationship is to give a referral, to give value. If you take every single prospect that comes to you, you don't have any leads to give to other people; you're just taking. When you niche, however, you're naturally saying yes to some things and no to many. Thus, niching allows you to give more referrals because every time you say no, that is an opportunity to develop a referral relationship with someone who can say yes.

This happens with my agency's clients all the time. Just as I was writing this, Daniel Hansen from Hansen Rosasco reached out to say, "Hey, we love your SEO and your agency is doing great on PPC, but we're really struggling with Facebook marketing."

We were able to say, "That's not a service we provide, but you should talk to (so-and-so) at (such-and-such agency). They're fantastic, and I think they'll really be able to help you."

Did *we* solve the problem? Nope, but the problem *will* get solved—and we get to be the ones making those connections. Daniel trusts us, so he trusts that the person we refer him to can solve his problem. In fact, he probably trusts us *more* because we showed that we have his best interests in mind. Someone else could have said, "Sure, we do that," and taken his money, knowing that they're not best suited to solve that problem. Because we were open and honest about what we do well and what we don't (which is just niching in a nutshell), we've reinforced the trust Daniel already had in our business.

Not only that, I'm in this niche so I already know the players on both sides: I understand the legal industry so I know what Daniel needs, and I'm in the marketing space so I know who can provide great social media services. He doesn't have to go research other people doing legal marketing, so he saves time.

There is certainly a *little* bit of effort and sacrifice on his part, in that he'll have a second marketing agency as a point-of-contact (versus just one if he were able to get that service from us). Having said that, do you really expect all agencies (or businesses) to be experts in *everything*? I would argue that an increase in expertise outweighs the minimal increase in effort.

Even if you are not the person delivering the end goal, by giving a referral you are still helping someone achieve their desired outcome. You are still delivering value.

THE BENEFITS OF REFERRALS

The best referral opportunities benefit three people: the person giving the referral, the prospect who is being referred, and the person receiving the referral. The prospect gets help solving their problem, the person receiving the referral will potentially get a new client, and the person giving the referral becomes a trusted advisor for the prospect and earns reciprocity with the person they refer to. This is one of those rare (but truly awesome) win-win-win situations.

Referrals create and build relationships, which can lead to marketing opportunities like speaking engagements, webinars, blogging, or podcasts. One of my top referral partners has a massive distribution and asked me, "Hey, can you write a blog on X?" Agreeing and writing the post meant that they not only promoted it on social media and listed my name as an author on their site, but I also got marketing distribution and an endorsement by this respected company in the legal vertical (my niche).

> An endorsement transfers trust from a trusted figure to the person or product being endorsed. We see this when LeBron James or Michael Jordan hold up a pair of Nike shoes. They're endorsing Nike. People trust that LeBron James and Michael Jordan know about shoes because they have been successful in their sport *in those shoes*. If they say that this is the shoe that they wear or recommend, then that trust transfers over from them to Nike.

Not only do referrals lead to better relationships, increased trust, and more opportunities, but they can also help you save money. A typical lead costs money in the form of both marketing and acquisition—but a referral is free. In fact, a referral is likely piggybacking off of money you already spent for a different lead. Once that person becomes a client, if they refer even just one person to you, that's twice the leads for the same money spent acquiring that first person. (But wherever it comes from—and we'll take a look at that in the next section—that referral is always free for you. Note: the moment that you *pay* for a referral, it's not a referral anymore; it's a lead.)

My largest clients have always been from referrals (typically from either clients or peers), and I've even found employees through them. When I have an open position, one of the first things I do is go to the people that work with my agency (my vendors and referral partners) and ask if they know anyone who is great in that position and who is looking for work. I almost always get a few strong candidates.

Ultimately, however, I think that the biggest benefit of both giving and receiving referrals is that you become a connector, and that's satisfying. It feels good knowing that you either got that person in front of the right person to help them, or that you are the person they were sent to, because now you get to help them. As human beings, we *like* to make connections. We like to help people. After all, that's why we do what we do.

You're building that goodwill currency. I don't know if you believe in the concept of karma, but the universe does seem to have a way of returning the positivity that you put into it.

THE THREE TYPES OF REFERRALS

By now you're probably thinking, "That's great, Chris, but *how do I get more referrals?*"

The number one way to *get* referrals is to *give* referrals. You've already seen the benefits of giving referrals, so let's take a look at the different types of referrals and talk about how you can earn more of each.

There are three categories of referrals:

- client referrals
- complementary service referrals
- competitor referrals

Each of the following sections covers one category of referrals in more detail.

CLIENT REFERRALS

First, you can get referrals from your clients.

There are a lot of tactics on how to get referrals (ask for them, have intentional processes and strategies, etc.), but one of the benefits of niching is that you'll be inherently referrable because you're an expert in what you do. If you've delivered someone's desired outcome for them, they're probably willing to refer other people who are like them and who want the same results.

It really is that simple: if you're good at what you do, it naturally lends itself to more referrals.

COMPLEMENTARY SERVICE REFERRALS

The second category of referrals is complementary service referrals, which refers to services your niche audience also needs but that you don't perform. I do a lot of marketing and lead generation, but that's only one component of a business. They may also need help with operations, sales, coaching and leadership, finances, and HR—those are all complementary services in my niche.

You can develop relationships with people who provide those complementary services in your niche so you can provide an end-to-end solution for your clients. Doing so saves your clients time, effort, and sacrifice, because they don't have to find all of those individual people on their own. It also helps them get the best service possible. Full-service agencies may sound appealing on the front end (because they have people that do everything), but because they don't specialize in an area, they may not be as good at some (or all) of the services they offer. When you niche, however, you truly specialize; you can then point to other people who are *also* experts in *their* area and your clients will get better results, all the way around.

That also cements your clients' decision to come to you (and refer you to others!) because not only do you know about your specific niche, but you clearly understand them, their pain points, and even their big-picture outcomes. You know the right people to help them solve problems along their entire journey.

With the other niche service providers, sending them a referral is the epitome of building a relationship. As business owners and entrepreneurs, our goals are to increase and grow our business, so if you directly contribute to those goals for another person, you are adding value for them.

When you first get started in your niche, you may understand that you should give referrals to other people, but not know whom to refer *to*. As you get to know the industry, a natural feedback loop occurs between your clients and who they also work with. You'll naturally learn who may be a potential referral partner for your business and your clients.

Take the time to fully develop those relationships. The greatest risk when making a referral is that the person you're referring to doesn't do a good job, which then reflects poorly on you. Your reputation is at stake, so be sure you know who you're teaming up with.

You may want to consider only giving referrals to other niched businesses. If you send clients to an all-things-to-everyone business for one of the services they need that you don't provide, that business can be a threat to you. For example, if I'm an SEO agency and I have a client that wants PPC, I'm not going to send them to Agency X, which does website design, SEO, and PPC, because they might end up poaching my client. My client also just isn't likely to get the best service from someone who doesn't specialize.

If, however, Agency Y is a PPC specialist and they only do PPC, it's symbiotic. I can send them my PPC clients and they can send me their SEO clients, with no threat to either of us.

Again, the greatest form of giving in business is *giving business*. Everyone wants a pipeline. Everyone wants to grow their business. If you send them referrals, they're going to want more referrals in the future. Chances are, if your service fits the needs of both of your clients, the people you've sent business to are more likely to refer additional people to you than to someone who's never given them any business at all.

Having said that, I subscribe to the mentality of giving without expecting anything in return. Ultimately, it's about helping your client to achieve their outcome.

> Not that you would ever do this, but be aware that there is a big difference between having referral partnerships and colluding to fix prices. Here's how you can tell them apart: referral partnerships create goodwill, and collusion is a crime.

COMPETITOR REFERRALS

The third type of referral is called a competitor referral, and it's exactly what it sounds like: sending referrals to your "competition."

Before you think I'm crazy or call for my head, there are some situations where this makes sense, such as when you have certain forms of exclusivity that prevent you from serving your exact prospects. When that happens, you can develop referral relationships with your direct competitors.

For example, I have a geographic exclusivity contract that prohibits me from working with more than one client in the city of Houston, Texas. I already have a client there, so when I get another Houston lead, I can't help them—but I *can* tell them who else may be able to serve them.

If you have exclusivity provisions in your niche, try to identify those businesses who have similar provisions and who are going to run into the same scenario. That way, if *they* already have a Houston client, but you don't, they can refer the lead to you.

You don't have to player-hate your competitors. You can be friendly with them and evaluate them objectively to determine who the best referral partners are. Referrals are an act of goodwill and truly helping your prospects. Even if you don't have exclusivity conflicts, sometimes it just doesn't work out with a client or they want to work with someone new. It's good to both know whom to send people to *and* to be the person your competitors would recommend in that situation.

When you choose a niche, you're choosing to have a reputation in that niche. The manner in which you treat referral situations has a long-term impact on your business.

Plus, if they know that you're friendly and send referrals their way, it just sets up a better relationship, where they're not trying to aggressively poach your clients or employees—and your competitors may even protect you. We have a referral partner who, anytime a client comes to test the water with them, says, "Oh, you're with Rankings? You're in good hands. Stay with them, talk to them. They'll help you figure it out."

Having that great relationship with referral partners can create a moat, a defensive situation. They're not ruthlessly trying to poach your clients. Instead, they're shoring you up and confirming to your clients that they made the right choice.

> I was recently listening to David C. Baker, a business consultant who wrote *The Business of Expertise* and who does a lot of speaking engagements. He asked, "What do you think the number one method is to get more speaking gigs?"
>
> The answer? The other people on the panel, who in many cases are your direct competitors. They likely get asked to speak at other events, and at some point in the future may suggest you join in too. "Oh, you should have Chris on, too. He was great on that panel last week."
>
> People expect competitors to speak poorly about each other, so when that doesn't happen—and then they even take the next step further and ask to invite the competition—it makes both parties look even better.

The reality is that not every agency or business is the same, and you likely won't be able to serve *every* prospect who comes your way. That doesn't mean one of your competitors can't! For example, our agency doesn't do Spanish content. Some of our competitors do though, so if that is something a prospect needs, we want to make sure they can still be helped.

Maybe you're just so busy that you can't help them right away. That's not a bad problem to have! In that case, you have two choices: offer to add the potential client to a waiting list, or give them a few names of people who may be able to help them right away.

Give Them Options

When referring clients or prospects to your competitors, you may want to make it *their* choice whom they ultimately work with. When you're unsure of who is the *best* choice for them, you want to give them options and let them choose for themselves.

If you refer a client or prospect to only one person, you're making the choice for them (and it's unethical if you're not entirely sure of the quality of service that they deliver). You're saying, "This is *the agency* you should use." When you give them several names, it's their choice to evaluate whom to work with.

This is important for several reasons. First, different agencies have different offers, and what's best for each individual is subjective and based on the prospect's specific needs. One may have long-term contracts, whereas another may have short term. One could be cheaper, another could be more expensive. If somebody doesn't have a big budget, they need a cheaper provider.

It's like when you and a friend or your significant other are trying to choose a place to go out to eat. You don't just give them one option; you ask, "Where do you want to eat? Do you want Italian? Mexican? Greek? Sushi?" They're all acceptable options. They'll all feed you, but what do you want? What do you need? What did you have for lunch?

> We can niche this up even further. If your significant other says they want Italian tonight, there is then a spectrum of Italian restaurants to consider. Do they want the romantic *Lady and the Tramp*' experience? Do they want Olive Garden's hospitaliano, or do they want real, authentic Italian food? Do they actually just want to order a pizza?

Let the prospect then do the due diligence to decide which business best suits their needs.

It puts the responsibility on whichever competitor the prospect chooses to then finish the job and win the business. If the client is choosing between two options, it's up to those two businesses to help the client determine the best fit. But *both* businesses can see that you put their name out there, which creates two forms of reciprocity as opposed to one.

That compounds your ability to get more referrals. If you just send all your referrals to one person, only one business is going to feel psychological reciprocity. If you send referrals to two or three people, then you have more opportunities to get referrals in return.

Look, I know the next paragraph is going to trigger a reaction in some readers…but I want to remind you that most of you started a business to *make money*. That's my warning. Here we go:

If you give only one referral partner your business and load them up on referrals, it could represent additional revenue for them, which can lead to them competing against you for market share in their marketing efforts for future new prospects. So another reason why you want to spread out your referrals to other businesses is to avoid building up threats in your market. They may be great people, but if they're only spending 10 percent of their marketing budget and you give them two million dollars in new business, guess what? They could now potentially afford to market *against* you.

ONLY BAD KIDS HATE SANTA

The issue most people complain about with referrals is that they are often feast or famine. They're unpredictable. That's true: they can be, because referrals are based on serving the needs of prospects.

That doesn't mean you can't tip the scales in your favor.

The best advice I can give you is if you want to get more referrals, be someone worth referring. Be a good person. Do a good job in your business. Be an expert, put in the time, niche up, and set yourself apart. (If you niche it, they will come!)

People get a sense of gratitude from helping other people. If you help people, you're valuable, and other people may want to help you because they'll get a sense of gratitude by doing so. When you give referrals and you're good to other people, you set yourself up to receive goodwill and collaboration in return.

That's why I say that only bad kids hate Santa—because the bad kids don't get gifts! They just get coal in their stocking. (Though with the price of coal going up, maybe that's not such a bad thing.)

If you are a good kid, you love Santa. Similarly, if you are a good person who provides a good service in a good niche, you will find referrals under the tree on Christmas morning.

We've looked at many advantages to niching up, but there's one more I want to discuss, and it ties everything together: Efficiency. Chapter 8 will show you how to create repeatable processes to save time and money.

CHAPTER 8

EFFICIENCY

Think about the very first time you drove a car. You were likely fifteen or sixteen years old or so, and everything just felt so *new and exciting*. That first trip, whether around the block or in an empty parking lot, probably required a lot of concentration: Where do your hands go on the wheel? What are your feet supposed to be doing? How far do you have to turn the steering wheel to move the car? How hard do you have to press the brake pedal to stop the car? Did you remember to check your mirrors? Whoa, you actually have to take one of your hands *off the wheel* in order to click the turn signal?! Is that safe??

Now think about the most recent time you drove a car. Did you consciously think about *any* of that? Probably not. When you've spent years behind the wheel, you get used to it. You understand how to drive the car without having to pay attention to the exact actions you take every step of the way.

What happened between those two car rides? Well, you've put in so many hours driving that you've more or less become an expert. Even more than that, you've done the *same* motions, over and over again, with the same results (allowing for minor changes as you get newer cars with new technology to make it even easier). Driving may have required some real effort when you were a beginner (and that's why your insurance cost more too, because inexperienced teenagers are more likely to get into accidents), but now you have mastered the technique. Not only have you become proficient, you've also become *efficient*.

> We can see this efficiency develop in other areas, too. Compare trying to roll your own sushi for the first time to the rolls you get from professional sushi chefs, who have performed the action tens of thousands of times. Consider a cake made by pastry chef Duff Goldman, who has made thousands of cakes, versus baking one for the very first time. Duff's is likely going to be better because he is an expert in his niche *and* because he has developed muscle memory from all that time spent in the kitchen.

BE EFFICIENT

We've talked about all the other pros of niching up that you've read so far, and this concept of efficiency ties them all together. You can be efficient with your processes, but you can also learn to be more efficient with referrals, with relationships, with conversions, and everything else just by gaining experience performing in one area with such expertise. By becoming more efficient, you can make more money, too.

When most people think about increasing profit, they think of raising prices and/or selling more products, but a large—often overlooked—aspect of profitability comes from eliminating waste. Being efficient can decrease your expenses. You can either increase your fees through your implied and real experience, or you can decrease your wasted time, which further increases your margins.

Niching up is a type of lean methodology where you exclude all other industries and just focus on one. By doing so, you also become more efficient because you eliminate redundancies and create repeatable actions.

Let's say that I try to reach a broad market by performing SEO services for professionals. I do keyword research for a personal injury law firm, then I have to do keyword research for home improvement, then for electricians, then physicians, and all the other different industries I market to. I have to recreate that research, from scratch, every single time for each of those different industries. Starting with a blank canvas means spending a lot of time for every new engagement.

On the other hand, if I just do personal injury law firm keyword research, I only start with a blank slate once. The next time I talk to a client who also needs keyword research, I've already done it, for the most part. There may be some small iterations to make, but I can utilize that initial work—and that initial time investment—again and again. It's repeatable.

We've talked previously about Henry Ford and his assembly line, but imagine if he tried to make every car different—different colors, different parts, assembled in different ways. How much slower would that be? From an efficiency perspective, how much time and effort would be wasted? If we look at what he actually did, creating one assembly line to crank out just the model T, he was able to preplan his strategy to be the most efficient and effective.

By niching up, you are able to preplan, too. You can research once and use that effort over and over again—without having to repeat the actual process. Because you work with a solution specific to your niche, that solution becomes more productized, allowing you to eliminate inefficient solutions that don't work for the specific group of people you're trying to help. Instead of having to create multiple landing pages, marketing messages, or sales processes for each industry you could possibly work with in a generalized approach, you are instead able to create one landing page, one marketing message, and one sales process specifically tailored to the smaller group of individuals you have chosen to work with—and whose pain points and desired outcomes you are able to understand on a deeper level because of that decision to niche. If you tried to work with ten different groups, you'd have to stay at a more surface-level knowledge, because you can't dive *that deep* with *that many* people.

Being efficient means using your resources without waste, and one of the most important resources we all have is time. Having the ability to build more efficiencies into your processes means saving more time. There are time-saving components to all areas of running a business, from marketing and attraction, to sales and acquisition, to delivery, finance, and administration. That time-saving is a form of value creation, which can be appended with a dollar amount. You're saving time, but you're also saving effort and energy, all of which means that you're saving money.

BECOME AN EXPERT IN EFFICIENCY

Efficiency also comes from experience. When you first get started, you're probably not going to be as efficient because you just don't know everything. Just like the first time you hammer a nail, you're going to miss it a little bit or it might not be straight. But after you do it enough times, you start to hit the nail right on the head. You gain speed, confidence, and quality in your actions. Because you're working in the same industry, you're going to have those repeated experiences over and over again in your niche, which leads to efficiencies and improvement in quality. By performing the same actions frequently, you reduce errors.

> Kaizen is a Japanese word meaning "continuous improvement." In lean manufacturing, it refers to the elimination of waste and redundancies. One of the best-known proponents of kaizen is Toyota.

In Chapter 3, we talked about niching leading to being an expert or gaining expertise in a specific area after putting in your ten thousand hours. By becoming more efficient, you eliminate waste in those ten thousand hours while adding more value to your time as you achieve expert status.

The biggest area of efficiency for my agency is in operations because we don't have to repeat keyword research. Keyword research is such a time-intensive aspect of search engine optimization: there are thousands of keywords for personal injury law firms alone. Imagine if I had to get thousands of keywords *per industry* across multiple verticals! How would it be possible to do that and be profitable? (Simple answer: it's not.)

But we don't have to recreate these processes from a blank canvas for every prospect who comes in the door. We know exactly where to get backlinks. We know exactly how to optimize their website. We know exactly what cases are worth from a content production strategy. At the end of the day, our personal injury agency is more profitable because we can refine and improve our editorial calendars and processes, repeating those we use most often so we avoid wasted time and duplicated work.

DISCOVER YOUR REPEATABLE PROCESSES

As you get more experience in your niche, start looking for what processes can be repeated because they are the same for all of your clients. You won't necessarily be able to see these areas of efficiency initially, but as you learn more about your industry and customers, you will discover the areas where you can save time.

Turn to your data. Track expenses in your books, profit and loss statements, or accounting software and see if there are places you can reduce waste. Review your operational processes for repeated delivery actions. Measure customer concentration by location and demographics. If you have a process for one part of your business or one demographic, it may be something you can use with another.

BE MORE STRATEGIC

Strategy means trying to forecast what will work, but a lot of that is almost like guessing based upon information that you have gathered through experience and analysis. It's not perfect.

But you can be more strategic by looking at the actions you've taken that have worked for one individual and seeing if you can apply those to more people.

Operations, delivery, or fulfillment is the largest area to consider because that's the car on the assembly line, getting prefabs and put together. The actions you take here to create the good or service are likely performed frequently, which means they will be the most resonant when it comes to efficiencies.

This can also be a place to consider human capabilities versus technology and robotics. Very precise actions that have to be repeated may be better performed using technological programs or robots. But also consider whether certain areas with repeated actions actually have to see those actions performed by you, the expert. Do they require that amount of talent? Are they extremely difficult? If so, you're likely still the best person for the job.

If not, this may be a place where you can hire someone who is not an expert in your niche but who can perform those actions competently so you can focus on the areas where your expertise *is* required. If everything is new and custom, you have to be (or hire) an expert with a wide range of skills. When you have repeated actions, even though you are perceived as an expert, you may not need expert technicians to perform those actions. A high school kid can quickly learn to pull the basket of fries out of the fryer when the timer goes off, whereas it may take years of experience to truly understand all the intricacies in properly putting sushi together. If you own a McDonald's, you can hire the less experienced person, but if you own a sushi restaurant, you likely want the chef who has already learned the skills and who has repeated the process countless times before.

Similarly, maybe there are five people who could build a landing page for your clients (which is a repeatable process), but you are the only one who has spent the time getting to know those clients, so your expertise is required to write about their pain points and your solution using their specific language. That aspect is not something that can be automated or handed off to someone without your skills and knowledge.

There is no shortcut to becoming an expert, but there are things you can do to free up some of your time and mental energy so you can better focus on those areas requiring your expertise.

IT'S ALL A MATTER OF TIME

The overarching benefit of being more efficient—and of niching overall—can boil down to one word: time. That's where value is created. That's also where profit is created. Time is what lends itself to trust, and a transaction is a form of time.

It's all time.

Saving time is powerful. Why? Because all of our time is finite, so it's the ultimate form of value. We're all going to die eventually, so the time we do have is precious and any moment of it that we can save is valuable. If we think of time as currency, niching allows you to save and manipulate time, which makes you more profitable and more powerful than somebody who is trying to do everything for everyone, *even if they're willing to devote more time to doing so.* An SEO agency trying to reach every attorney, physician, and accountant will have to spend so much more time, but will still be so much less effective, than the one just focused on the legal industry.

Even if that generalist is willing to spend the time, is it worth it? They're not going to be as effective because they don't have the expertise, so they're not going to be able to charge a premium, they're not going to convert their prospects as easily (or at the same rates), and they're not going to build up the same amount of relationship equity (so they aren't positioning themselves to give and receive referrals either).

You, however, now know the benefits of niching up so you are ready to save time, be more efficient, and make more profit.

CONCLUSION

Now that you have read about the pros and cons of niching up, I hope you have a clearer picture of just how much those advantages outweigh the disadvantages. But I want you to know that I'm not just writing about niching; I'm actually doing this too, every single day. Let me be clear: I'm not the best at *many, many* things, all of which my wife would be happy to detail. Having said that, I know that niching up has created my own area of expertise, a space where I feel confident enough to tangle with the best.

I want to use my business as a case study, to show you how I niched up and discovered each of these advantages as I went from attorneyrankings.org to rankings.io—not just a marketing agency, or even a law firm marketing agency, but a *personal injury law firm marketing agency*. Let me walk through each of the pros and cons to show you how I made the decision about where to niche, and why that decision was the right one for me.

First the so-called cons, which as you've seen are not always a bad thing:

Smaller Market: I found that 70 percent of my revenue was coming from less than 40 percent of our clientele. Even though our market cap was smaller, that data told me there was a large enough market to support a niched business. Even though this *can* be a con, my profit was showing it wasn't a true con for my niche because there was a large enough opportunity to support it.

Waste: Niching itself actually allowed us to eliminate waste. I didn't offer every digital marketing service; I eliminated the ones that don't help PI law firms and just kept the ones that do, which also meant I didn't need the same staffing capabilities. I was similarly able to eliminate unnecessary expenses in marketing—instead of marketing everywhere and throwing paint against the wall, I'm essentially throwing paint right at PI attorneys—as well as in the sales process, operations, billing, and many areas of our business. (More on that in the section about Efficiencies!)

Competition: In my industry, and especially with the mobile devices we all have in our pockets that allow us to connect with anyone anywhere, clients can choose any digital advertising/marketing agency in the world to work with. But people tend to trust individuals that are similar to themselves, so by focusing on our clients and trying to provide value to that interest group, we've found a way to differentiate ourselves and stand out from the competition. We are the only digital agency that I'm aware of at this time that does personal injury marketing that isn't a heavy media buyer (doing traditional advertising on TV and radio).

Lack of Diversity: I've mentioned this before, but I don't even see this as a con. I have a passion for SEO. To me, it's like playing a video game that pays me. In video games, you're always trying to level up, get more experience, and make more money. I'm doing the same thing, just implementing it in real life. Just like in video games, where you have different obstacles, it's how you overcome it to grow. SEO is a zero-sum game; there's only one ranking number-one position. I enjoy playing to win.

Industry Risks: That doesn't mean that I'm not aware of potential risks of being in this industry. Even though attorneys have been around forever, and individuals are always getting injured (thus necessitating personal injury attorneys, who then need marketing), industry risks are real for my agency. Until people stop being injured, there will be a space for my agency, which allows me to insulate myself somewhat to that risk. However, this is still a con to be aware of. For example, social media marketing used to just focus on Facebook, Twitter, and LinkedIn. Times and attention spans have changed to YouTube, TikTok, and Instagram, etc. Because I am in the digital media game, I can see the shifts coming and am in the position to make a decision to quickly move.

Product Perfection: Because we only work with personal injury attorneys, we are able to survey our clients about their experience and continually optimize for those specific individuals. If I were a digital agency that worked with a thousand different companies across different industries, they would all have a thousand different needs. I get an amazing amount of feedback, which allows me to get better at doing the same thing, making small incremental improvements to continually widen the gap and be the best.

Effort and Sacrifice for the Buyer: Because I'm aware that, from an effort and sacrifice standpoint, we don't offer every type of marketing service, my agency is proactive. We partner up with the best pay-per-click legal marketing agencies, the best media buyers, the best in sales and operations. By knowing our clients, we understand their needs and are able to then find other niche companies, other experts in their craft, to help them. Because I can't help my prospects with everything, I can also give out referrals and set up mutually beneficial relationships. The value in quality and referral relationships outweigh the increased effort and sacrifice in having to talk to a couple different agencies.

When you niche, you can actually turn some of those "cons" to your advantage. Now let's take a look at the pros of niching up:

Awareness: I had to have a lot of experiences to understand where I had competency and where there was demand and opportunity before I niched up into my current industry. I did a lot of affiliate marketing, with more than 100 niche sites. Then I worked for a digital agency that worked primarily with legal but also had other types of clients in HVAC, home services, etc. From that experience, I understood that I worked well with legal, so my first niche was legal marketing. I could have offered digital marketing services to everyone, because I knew where to advertise for attorneys, but I wanted to use my time and capital in the best manner, where the biggest opportunities were. After seven years, when I had enough data to make the decision and was aware of this opportunity to niche up further, I focused in on SEO for personal injury law firms.

Additionally, finding that niche allowed me to learn the specific needs and nuances of that smaller market. Awareness can mean many things, but for our agency it means understanding which digital marketing services help personal injury law firms. There are a lot of marketing tactics, and I am an omnichannel proponent—having said that, certain marketing channels outperform others. We are aware of those that have the highest likelihood of driving leads or improving their brand recognition—but we wouldn't have that awareness without working within a *specific* area of the law.

Expertise: In our industry, pay-per-click tends to be more difficult than in other industries because of the cost per click. A car accident lawyer could pay several hundred dollars per click; it's highly competitive. In order to do television marketing effectively, you have to saturate the market and be one of the top three advertisers—in most markets, that's a six-figure investment. Depending on the size of a personal injury law firm, that may not be conducive to them. We understand where a firm is and what the best services are for them. We can concentrate on providing more value and going deeper with that relationship instead of just guessing, trying to offer everything to see who will buy it. We know what our clients want and need, and we can cater to those needs.

We do SEO every single day. We're talking to personal injury law firms every single day—and we're not distracted by other services or industries. We specialize in search engine optimization. We have SEO curriculum specific to legal. We're SEO masterminds. We have an external SEO coach, SEO speakers. We monitor Google's SEO news on a daily basis. Our entire presentation is continually refined by what's new in SEO. We get to see trends. When you have an entire company dedicated to doing one thing for one market, it lends itself to innovation and true expertise.

Premium Pricing: Most SEO agencies don't understand the quantitative actions needed to get results for personal injury law firms—they don't have the awareness because they've never competed in the industry, so they dramatically underestimate what is needed to get results in a highly competitive market. We know from our experience working with hundreds of PI firms what it actually takes in each market. We can evaluate it on a deeper level. That's why our pricing is higher—not because I just chose to charge high fees. It's because the market demands it.

This speaks to another point about my industry specifically: SEO agencies frequently get a bad rap as snake oil salesmen, and maybe I'm being too generous in my analysis, but I don't think there are very many people in my industry who set out to be deceptive. Rather, I'm of the opinion that the truly competent agencies in this vertical do it so well that an outsider can misinterpret it as being easy to do. Those folks tagged as shysters come into the SEO space with every intention of helping their clients grow, but unfortunately, they aren't adequately prepared for what's required to succeed here.

Conversions: When we speak to a prospect, we're not speaking in generalities of law firms; we're speaking specifically to them and their pains. We understand that a PI attorney typically wants more motor-vehicle accident cases, and we understand the value of those kinds of cases. We understand what keywords they need, the types of content they want. When we talk to them, we are able to mirror our clients and the questions they ask from a deeper level, which lends itself to trust.

Because this is the niche we focus on, we know who they are: I know that most of our prospects are male, in their forties to fifties, former athletes who have competed in sports. Most of them are trial attorneys who are very competitive; they want to win. I understand their desire to win and be the best—and the way I speak to them is completely different from how I would talk to a family law attorney, whose clients are going through a difficult time and have an entirely different persona. I can cater the experience specifically to the PI attorney I speak to—for example, if they are the owner, we focus on the big picture outcome as opposed to the granular details a marketing manager may want.

Additionally, all of our case studies and testimonials are from PI law firms. Our podcast is the *Personal Injury Mastermind*. I know this world and the people in it.

Relationship Equity: Most people think about relationship equity from a financial perspective, but I think about how relationships scale when you choose an industry. For example, when I was getting started in my industry, I worked with solo practitioners and small personal injury law firms. I certainly didn't have the attention of Glen Lerner, Kyle Bachus, Darryl Isaacs, or Mike Morse to invite them to be on my podcast—they had no idea who I was! But as I started to get results and move up the ladder, I was introduced to more and more personal injury attorneys and their friends. Now, for the most part, if I want an introduction to a specific person, someone in my network knows that person and can help get me that introduction. It just scales.

For example, look at the back cover of this book—you'll see blurbs from Joe Fried and Mike Papantonio. Joe Fried is one of the top trucking attorneys in the country, has litigated cases in more than thirty-five states, and co-founded the Academy of Truck Accident Attorneys. Mike Papantonio has written multiple books, made appearances in both film and television, hosts a major conference, and has been inducted into the Trial Lawyer Hall of Fame. Their time is precious, and without that network of relationships I've built up, there's no way they would have come on my podcast or agreed to blurb this book.

Referrals: As of the writing of this book, my agency, rankings.io, offers two services: website design and search engine optimization. That means that we choose to say no to providing every other service personal injury attorneys ask for. Through that focus, we have identified the best specialists for those services, because we still want to provide value to our clients by referring them to the best. We refer to another niched agency for pay-per-click—and that agency doesn't do SEO, so they send referrals back to us. We also refer out video production, coaching, and sales intake. We try to identify and continually improve our referral partners, to help our clients solve the pain points we can't solve for them.

Not only that, when you choose a niche, you become known for that niche. So if another digital agency has a personal injury law firm as a prospect and that agency, for whatever reason, doesn't want to or is unable to help that prospect—if that prospect types in "personal injury marketing," they're going to find us because we've built a brand around this niche, which is a unique selling proposition.

Efficiency: We focus primarily on SEO, ranking a law firm's website on the first page of Google. One of the most difficult and time-consuming activities that an SEO company has to do is keyword research, to determine what the law firm's clients are typing in as search terms to try to find the law firm. Because we focused on PI, we knew that most PI firms wanted the same types of cases so we could create our keyword research once and not have to do it from scratch again. This allowed us to eliminate wasteful activities that weren't repeated very frequently. I could do it once, continually improve it, and have a much better product.

And that's just one example! We have built repeatable processes into every area of our business: keyword research, content strategies, acquiring backlinks for personal injury law firms, creating landing pages, advertising (I know which are the biggest markets and where to advertise for PI law firms), copywriting—even email marketing. Most digital marketing agencies will need to use some logic-based email platform to segment their audience, but because my audience is all personal injury law firms, sure there may be different sizes and price points, but we can speak to one interest group and we can use a simple newsletter to fulfill most of our requirements.

NOW IT'S YOUR TIME TO SHINE

You've read this entire book and learned from my stories—now *you* are ready to niche up and discover the advantages for yourself: greater awareness, better conversions, more freedom, more time, more profitability, a greater reputation, improved ability to deliver results, and more valuable relationships.

We've talked about wealth and status, and niching up gives you both. It leads to greater profit margins, and you also improve your status because you're better at what you do, you are an expert in the space, and you're confident in the results you deliver.

Niching up gives you a sense of gratification, a sense of confidence, and a feeling of goodwill, knowing that you can truly serve the people that come to you, who need your help.

It was my goal, in writing this book, to change the narrative on niching from being about scarcity to having an abundance mindset. Niching is not restrictive; it provides *opportunity*.

In a world of saturation and competition, all of us just want the best. When you niche, you focus all your efforts on doing what is best to help your customers. If you're a business owner and you have a passion for helping a certain market, I would encourage you to go all-in—to think not about the cons of having less, but to look at all the advantages of what niching up can mean for you, from the perspective of wealth and happiness.

If you find yourself competing in a red ocean with a ton of competition, you may benefit from looking at your audience and seeing who you can truly help provide the most value to. Find your own blue ocean—and remember: the narrower the market, the bigger the prize.

If you want to see how I've implemented these niching principles, go to rankings.io.

Follow me on Instagram (@chrisdreyerco) and tag me to tell me how you niche up.

If you're a personal injury attorney, I have another book coming out specific to your interests. In the meanwhile, if you want to hear more about what I stand for, check out my *Personal Injury Mastermind* podcast.

ACKNOWLEDGMENTS

I would like to thank:

- Jenna, for allowing me the freedom to use my OCD as a superpower
- Dad, for always teaching me to play to win
- Mom, for always having my back, even in those rebellious days
- Alicia, for giving me the $15k bootstrap to pursue my dream
- Matt, for helping me find my voice in this book
- Jenny, for pulling it all together
- Ed Dale, for his 30-Day Challenge that helped me learn digital marketing
- The principals at Herrin High School, for giving me the leeway to learn internet marketing in the detention room
- My detention room delinquents, for giving me the push out the door

ABOUT THE AUTHOR

CHRIS DREYER is the CEO and Founder of Rankings.io, an SEO agency that helps elite personal injury law firms land serious injury and auto accident cases through Google's organic search results. His company has the distinction of making the Inc. 5000 list five years in a row.

Chris's journey in legal marketing has been a saga, to say the least. A world-ranked collectible card game player in his youth, Chris began his "grown-up" career with a History Education degree and landed a job out of college as a detention room supervisor. The surplus of free time in that job allowed him to develop a side hustle in affiliate marketing, where (at his apex) he managed over 100 affiliate sites simultaneously, allowing him to turn his side gig into a full-time one. When his time in affiliate marketing came to an end, he segued into SEO for attorneys, while also having time to become a top-ranked online poker player.

In addition to owning and operating Rankings, Chris is a real estate investor and podcast host, as well as a member of the Forbes Agency Council, the Rolling Stone Culture Council, Business Journals Leadership Trust, Fast Company Executive Board, and Newsweek Expert Forum.

CPSIA information can be obtained
at www.ICGtesting.com
Printed in the USA
LVHW040047270523
748224LV00010B/145/J

EPIDEMICS

EPIDEMICS

AN INTRODUCTION

Dr. Kiumarss Nasseri
Dr. Paul Mills

Copyright © 2022 by Dr. Kiumarss Nasseri Dr. Paul Mills.

Library of Congress Control Number:		2022907749
ISBN:	Hardcover	978-1-6698-1725-3
	Softcover	978-1-6698-1724-6
	eBook	978-1-6698-1726-0

All rights reserved. No part of this book may be reproduced or transmitted in any form or by any means, electronic or mechanical, including photocopying, recording, or by any information storage and retrieval system, without permission in writing from the copyright owner.

Print information available on the last page.

Rev. date: 08/26/2022

To order additional copies of this book, contact:
Xlibris
844-714-8691
www.Xlibris.com
Orders@Xlibris.com

Epidemics

WHAT THEY ARE

WHY THEY APPEAR

HOW TO DEAL WITH THEM

CONTENTS

Preface ix

1. **What is an epidemic?** 1

2. **Definition** 5
   Epidemiology and Public Health 7

3. **Basic Elements** 11
   The Cases 11
   Iceberg Phenomena 15
   The Population 16
   Birth 17
   Death 18
   Migration 19

4. **Theoretical Construct** 21
   The Agent 22
   The Host 24
   The Environment 27

5. **Patterns of Occurrence** 33
   General Distribution 33
   Specific Distribution 36

6. **Disease Accounting** 39

7. **Types of Measures** 47
   Statistical and Clinical Significance 49
   Association and Causation 51

8. **Research in Epidemiology** 53
   A. Study Question 53
   B. Study Population 53
   C. Required Data 54

    The sources of Data 55
    The Total Count 55
    Sampling 57
    Sample Size 57
    Sampling Frame 58
    Sample Selection 59

9. **Data Handling** 61
    Data Analysis 61
    Data Presentation 61

10. **Types of Epidemiological Studies** 63
    Study Direction 63
    Study Classification 65
    Relative Risk 69
    Experimental Studies 70
    Phases of Randomized Clinical trials 71
    Placebo Control 73
    Informed Consent 74
    Clinical Trial, Trial and Error 74

11. **Prevention** 77
    Level 1. Environmental Development 78
    Herd Immunity 79
    Level 2. Personal Protection 81
    Level 3. Diagnosis and Treatment 82
    Screening and Early Detection 83
    Level 4. Rehabilitation 85
    Level 5. Palliative Care. 85
    Eradication and Elimination 86

References 89
Index 93

PREFACE

The pandemic of COVID-19 has significantly increased the usage of words like outbreak, epidemic, and epidemiological study in daily life. If you have ever wondered what an epidemic is, why it happens, and how it is controlled, or have ever struggled to follow an epidemiological study and find out why and how it is done, then this brief book is for you. Epidemics and epidemiological studies are major issues discussed and explained in the academic discipline of epidemiology, which is the engine behind public health and community well-being. In a nutshell, epidemiology is a scientific method that begins with the description of the distribution of diseases in a community, their mass occurrence as an epidemic, sets out to find the reasons for its occurrence, and then suggests methods for dealing with it. Epidemiology provides the road map, and to reach the destination, one should be able to read and understand the map. Public health is not something that happens to people, but *it* is a process in which people are a major partner. A well-described and clearly-explained knowledge of the essential concepts and terminology in epidemiology helps with the realization of the extent and nature of disease problems in the community,

We have rearranged the shelving. Politics are next to comics, travel guides are in fantasy section, and epidemiology is down the hall in "Current affairs".

and thereby enhances public participation in resolving it. Elevated public understanding of the basics of epidemiology is a significant contributor to the advancement of public health activities as well as improvement and maintenance of the general well-being of the community. This booklet aims to provide such an understanding for the interested public. Numerous academic texts and scientific articles are prepared on various aspects of epidemiology with extensive technical detail and are regularly used around the globe at various institutions of higher learning. The material included in this booklet provides a brief description of the basic concepts and methods of this science, in a simple and innovative context for the public. It can also be considered an introduction for those who fancy this branch of science as a professional career. Material presented in this booklet is, by necessity, brief and void of detailed and elaborate technicalities or description of specific diseases. Nevertheless, relevant references to major points and concepts are also provided for interested readers for further information.

1 | WHAT IS AN EPIDEMIC?

"Epidemic" is a general label used for pointing to an unexpected occurrence of a disease or event in a population or community. The earliest mention of the term epidemic is in the Holy Bible, Book of Exodus, as various plagues were imposed by the Lord on Pharaoh Ramses II. These included swarms of frogs and locusts, as well as widespread human diseases and afflictions of various kinds.[1] For a long time, the mass occurrence of diseases in human populations was considered punishment by God, and it was not until the discovery of microorganisms that its biological nature was identified. The word "epidemic" was first used in early-seventeenth century in France based on its Latin origin and is widely used for the description of the mass appearance of severe and deadly diseases like black death and cholera, which had significant public health impacts in the community. Since the Spanish Flu in 1918, a number of major epidemics like Asian Flu, AIDS, Swine Flu, Ebola, SARS (severe acute respiratory syndrome), and Zika have claimed the life of many millions of people.[2] The word and concept of epidemic has also been used to characterize the rapid increase of disorders with major socio-behavioral elements, such as the spread of gun violence,[3] misinformation,[4] and xenophobic behavior.[5]

Currently, COVID-19 has engulfed the globe in a pandemic which is defined as the occurrence of the same disease in epidemic proportions across national borders in many countries. With over one million deaths in the US, COVID-19 has significantly surpassed the impacts of Spanish Flu epidemic of 1918-19 that caused about 675 thousand deaths. Globally, COVID-19 has affected close to 600 million individuals with over 6 million deaths since its appearance in January 2020 in China[6], is not contained yet,[7] and is expected to stay with us for years to come. Unfortunately, it will not be the last pandemic. With significant damage to the environment and the disturbance of the delicate balance of the ecosystem, deadlier epidemics are not only probable, but inevitable in the coming years.

The occurrence of epidemic disease in a community is a signal that something has gone wrong that needs attention. It is similar to the occurrence of fever or pain in an individual that points to some health problem requiring closer attention. Epidemic in this sense is a signal that disease has spread in the community, and its detection, management, and control need expertise and knowledge beyond that of the clinicians treating the disease in individual patients. Major characteristics of epidemics include their unusual appearance beyond what is normally expected and the fact that they do not happen by chance. There is always an extra ordinary reason or sets of reasons for epidemics to happen, and once they happen, their containment and resolution also require extraordinary effort and resources. Identification and understanding the reasons behind the appearance of an epidemic is a must for selection of the actions that can be taken to resolve it and preventing it from happening again in the future. Dealing with an epidemic requires a combined effort and collaboration among health professionals who identify the problem, civil administrators who design and conduct control measures, and the public who follow the

I am sorry, we are epidemiologists not dermatologist. But, we can see you if you manage to get a group of you with the same problem.

recommendations. Epidemiologists are part of the health professional team who identify and document the presence of epidemics, provide scientific support for control measures, monitor the progress, and evaluate the outcome of containment efforts. Detailed explanations of epidemics and their control methods are well explained in the science of epidemiology.

Public health authorities, experts in medicine, mass media anchors, and many other members of the community excessively use the terms epidemic and epidemiology as talking points in public forums and interviews. Despite this widespread use, large segments of the general population who are exposed to this terminology in daily conversation may not have a clear understanding of what the term means and what it represents. For many, epidemic is a fearful word pointing to severe problems like wide spread occurrence of common diseases or bad behavior in the community. For few others, it is simply Greek jargon similar to other branches of medicine. Still, for others, it is a black box that provides evidence and support for all sorts of statements by physicians, scientists, and reporters. In reality, epidemiology is an important field of medicine and is the basic tool in public health for understanding the dynamics of mass diseases in a community and finding ways to confront them.

2 — DEFINITION

Epidemiology is the study of diseases in the community: who gets it, why they get it, and how to deal with it. In other words, it is the science and art of determining what is happening to the people in a community and why, with the eventual goal of suggesting ways for desired intervention. It is based on the premise that mass occurrence of events in a population does not just happen by chance, but follows specific patterns of causation and distribution. Hence, the objective of epidemiology is to describe the current distribution of the event, determine what happened in the past that caused it, and suggest what should be done to intervene and prevent its occurrence in the future. In other words, the the ultimate objective of epidemiology is to control the spread and promote the elimination of disease in man. Epidemiologic principles can also be used for support and expansion in the community of desirable events such as vaccination, good nutrition, and healthy ways of life. Modern epidemiology was originally defined as the scientific method for the "study of the distribution and determinants of disease."[8] This definition has expanded over time to cover all health-related events: "the study of the occurrence and distribution of health-related states or events in a specific population"[9] and is considered the basic scientific tool in the practice of public health. Epidemiology is a scientific research method and, as a deductive science, is based on factual data that can be observed, measured, and assessed for accuracy and reliability. Conjectures have no place in this field of study. Moreover, it is always more than the simple sum of its accumulated facts. The added value comes from the analysis and utilization of the information that is extracted from data.

> *Epidemiology is like a box of chocolate. It looks exciting on the outside, but the essential matter is inside.*

In practice, epidemiology begins with describing the current situation, searching for the past events leading to it, and suggests methods for proper intervention. Then it continues with the assessment of the effectiveness of the adopted interventions by the official authorities. In short, epidemiology studies the present (what) and searches the past (why) to find ways (how) to intervene in the future. It is the science of the mass occurrence of events and although originated in disease control and public health, its concepts, principles, and methods can effectively be used for the study of any mass phenomena. When dealing with disease and epidemics, the objective is control, prevention, elimination, and eradication; when dealing with favorable behavior like smoking cessation, vaccination, or charity donation, the objective is support, promotion, and expansion; and when dealing with community services and established interventions, the objective is identification of the bottlenecks and shortcomings to suggest improvement and enhancement of effective interventions.

Since the concepts of epidemiology can easily be applied to any mass phenomena, an expanded definition for epidemiology can be stated as "the science and the art of mass phenomena." This definition underscores three major components of the discipline: the science, the art, and the mass phenomena. The science of epidemiology is a well-defined set of principles, methods, and applications that are used in dealing with facts about the phenomena under study. Numerous institutions and organizations provide this kind of education in various shapes and forms at different levels of training, from field staff to advanced researchers. The art of epidemiology, on the other hand, is mainly experiential and cannot be taught in a formal manner. It is gained by practical laboratory or field experience, depth and breadth of knowledge about the phenomena under study, and familiarity with the social and cultural environment of the community of its occurrence. The same set of data in the hands of two epidemiologists with different experiences and outlook may result in recommendations that can be different in substance and approach. The principles of epidemiology are the science and its application is its art. Finally, a mass phenomenon refers to the occurrence of a particular event in a large segment of the population and, in obvious excess or deficit of what would normally be expected to occur.

Epidemiology is a collaborative discipline and cannot be practiced as a stand-alone profession. Unlike many other professionals who can practice their craft individually and privately, epidemiologists are required to work along other technical and administrative professionals who are striving to achieve the same objective. As members of the academia and research institutions, they search to expand the frontiers of the science and train future epidemiologists. As members of administrative bodies, they provide the blueprints for quelling epidemics, improving the impacts of population issues, and modifying the outcome of interventions in desired direction.

Epidemiology and Public Health

Although epidemiology can be considered as the "engine" of public health, it is not the "driver." Public health or community health covers the well-being of every single citizen in a given community and is based on three major pillars of scientific knowledge, official mandates, and community participation. Scientific knowledge is enhanced by medical practitioners and researchers concerned with the nature and treatment of the disease; administrative actions implemented by the government and other official administrators contribute in terms of providing rules and regulations; and the people in the community contributes by observing the suggested healthy way of life, such as refraining from hazardous materials, maintaining good nutrition, and practicing a self-preserving mode of life. Epidemiology is the background science that contribute by describing the distribution of the disease and reasons for it; suggests control methods, and helps with evaluation of the effectiveness of the administrative planning and community participation. As an example, epidemiologists identify that measles is a disease of childhood which confer long-lasting immunity, physicians and other scientists develop the vaccine, the government passes laws and regulations requiring vaccination of all children, and the public participates by following the rules and vaccinating the children. Epidemiologists monitor the progress of the campaign, which at one point had resulted in the elimination of measles in the United States. Now it has partially returned due to refusal of some parents to vaccinate their children. If any of these aspects do not materialize, the public health of the community from the standpoint of this disease will be in jeopardy.

Epidemiologists function as scientific and nonpolitical advisors to decision makers and are trusted to provide evidence-based information on various aspects of the issue at hand; they suggest and support intervention plans and monitor and evaluate the progress and achievements of the adopted plans. As an example, in the fight against lung cancer, the causal connection between the disease and smoking was discovered and described by epidemiologists, but actions toward modification of smoking habits that aimed at significant reduction of lung cancer was entirely legal and administrative in nature, and most important, it was the public who actually stopped smoking. Finally, it was incumbent on epidemiologists to monitor and document the progress of these interventions, Figure 1.[10]

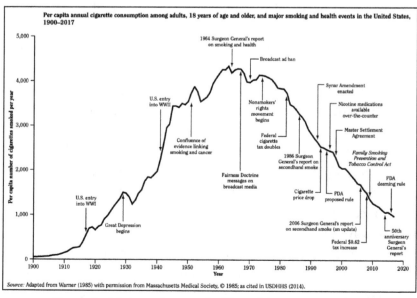

Figure 1. Adult smoking in the United States, 1900–2020

Although the principal focus of epidemiological studies is mass phenomenon, *i.e.*, diseases, events, or other nefarious circumstances in populations, its ultimate goal is enhancement of the well-being of human life through development, monitoring, and improvement of effective interventional measures. The practice of epidemiology began in the field of disease control and public health but has presently evolved into a distinct discipline of population science that can effectively be applied to the study of any mass phenomena, including those in the domains of social justice, such as racism, migration, law enforcement, as well as economic behaviors like effective advertising, marketing, and other issues that involve large segments of the population.

3 BASIC ELEMENTS

There are two principal elements in epidemiology: the cases of the disease or the event of interest (or the numerator) and the populations at risk of developing the case (or the denominator). Cases define the disease or the mass event of interest, and the population defines the context of where and when it happens.

The Cases

Case is another name for those individuals who are affected with a disease or an event of interest. Identification of cases is probably the most significant undertaking in epidemiology that requires a clear and practical definition. While detecting the occurrence of a disease or an event in a single individual is relatively easy and is generally dichotomized into a "yes" or "no" conclusion, detection of cases in a population is complex and not so easy. The reason is that the same disease may have different clinical presentations in different individuals. The objective of case definition is to identify all cases of the same disease in the population; however, considering the variety of signs and symptoms of the disease, this is not a simple task. To collect reasonably-complete data on all cases in a population, a clear, comprehensive, and practical case definition must be developed. The major points in case definition are the specificity of the sign or symptom in identifying the problem, the frequency of its occurrence in those affected, and its measurability in a uniform and concise manner. Case definition is an important step and must ensure that all those who are identified for further study share the same disease or issue. Moreover, case definition should effectively represent the largest possible group of affected individuals who may have different expressions of the same disease. Variability of the symptoms in different members of the population affected with the same issue is normal and common in many diseases. Care must be taken to identify as many people affected with the same condition, even though they may represent different expressions, as possible in the case group. It is crucial to not mix individuals affected with a different disease or issue who may present with similar signs and symptoms. Mixing cases of diseases or events in itself is a major source of error that requires diligent attention. As shown in Figure 2, while the population and the mass event have irregular boundaries, the criteria, A–F, selected for case identification is generally more specific with clear boundaries.

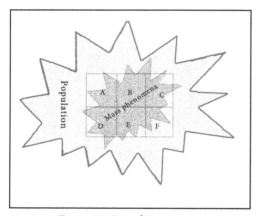

Figure 2. Population, mass phenomena, and determinants

Close examination of this figure reveals that some of the specific criteria selected to identify cases can be observed in both affected and not affected individuals. Thus, an important step in any epidemiological work is to produce a reasonable set of criteria that are easy to use and specific enough to maximize the selection of cases of interest and exclusion of "non-cases." The critical concern in reporting of cases is clarity about the criteria used for case identification. One cannot draw meaningful conclusions about the nature of the apples or oranges when they are mixed in the same study.

As an example of case definition, consider a disease like COVID-19. Multiple symptoms and laboratory findings are associated with it and include chest pain (76%), fatigue (69%), cough (63%), headache (60%), body ache (58%), fever (57%), loss of either smell or taste or both (56%), and few other symptoms.[11] Almost all these signs are also associated with other diseases. Thus, relying on any one of them is not sufficient for definite diagnosis. Additionally, two laboratory tests: reverse transcription polymerase chain reaction (RT-PCR), and rapid antigen tests that measure different aspects of the disease. have also been developed and are in wide use. Computerized tomography (CT Scan)), are also available for confirmation of diagnosis. The RT-PCR test, which takes a couple of days to complete and is considered the gold standard for diagnosis of this disease, is generally positive in only 71% of cases that are identified in other ways, and the CT scan of chest can identify up to 98% of cases in the hospital.[12] It is important to determine which of the various symptoms by themselves or in combination with others is selected as the definitive and accurate criteria for the diagnosis of this disease. Moreover, the criteria selected for case identification must be measurable and relatively easy to conduct under different circumstances in affected individuals with the same results.

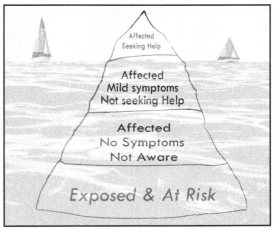

Figure3. Iceberg phenomena

Another important point in case definition is its utility in covering and detecting all or a large portion of affected individuals in the population under study. Affected individuals present with different symptoms, are seen at different facilities, are diagnosed based on different criteria, and are widespread geographically and chronologically. Laboratory tests are not similarly available in all localities, the diagnostic signs or symptoms are not similarly detectable in all cases, and people have different patterns of requesting medical attention. Moreover, many of the symptoms are shared with more than one disease. Thus, for accurate identification of all individuals in a population with the same disease or entity, a set of criteria must be developed that are highly specific, relatively simple, and easy to conduct. These criteria must be generally attainable in an extended geographic area and applicable to a large segment of the population under observation.

For some events, like common infectious diseases, detection and identification is rather simple and straightforward based on a combination of clinical symptoms and laboratory test results. However, even for well-known diseases, like pneumonia, different individuals may exhibit different symptoms and test results. Development of a set of clear and objective definitions and standardized detection criteria will ensure that those identified as experiencing a given event are actually affected with the target condition. It is incumbent on the epidemiologist and is probably one of their highly-critical missions to arrive at the clear and workable diagnostic criteria for inclusion of cases and exclusion of non-cases. Some of the approaches like the use of HbA1c[13] for detecting diabetes and body mass index (BMI) for detection of obesity are straightforward, based on statistical analysis of large population studies, and are easy to detect and measure. These diagnostic criteria, as well as specific methods for its observation and recording, must be developed and standardized in collaboration with professionals who are specialists in the disease or the mass phenomenon under study. Disease surveillance organizations have developed case definition lists that are open to the public and are regularly updated as new information is collected. In the United States, the Centers for Disease Control and Prevention (CDC) has created and posted case definition lists for most diseases.[14] A major advantage of such lists is creating comparability for diagnoses made on different occasions and, thus, help with comparing or combining studies on the same disease which are completed at different locations and times. However, for complex concepts like some chronic diseases, most social behaviors, and other mass phenomena, arrival at specific detection criteria and identification methods are not easy.

Another major concern with identification of the true extent of an event in a population revolves around the fact that mild cases are generally not reported to the authorities. This is referred to as the "iceberg phenomena" that mainly include the most obvious cases. Obviously, diagnostic and reporting capabilities of diseases may not be equal in all locations, and a diagnosis by a healthcare professional in the clinic or hospital setting is more concise and accurate compared to a self-diagnosis by respondents in a mail survey.

Iceberg Phenomena

Iceberg phenomena refers to the detectable and reportable segment of the disease in a community, that is easily noticed. Surveillance which is monitoring the presence and extent of the disease in the community, actively looks for the icebergs of unusual or serious diseases. The main concern with this activity is that a much larger number of cases of a disease or mass phenomena may be hidden in the community, Figure 3. In epidemiology, the few cases that are noticed by authorities are referred to as the tip of the iceberg, triggering efforts to search and identify the extent of the disease that are hidden in the community. As an example, detection of few cases of Kaposi sarcoma and pneumocystis pneumonia in Los Angeles in 1981 was the tip of the iceberg for the CDC epidemiologists, whose further investigations resulted in identification of AIDS and HIV infection.

Detection of the tip of the iceberg for social behavior, such as racism, domestic violence, adultery, or negligent driving, are generally based on cases reported to the respective public health and law enforcing officials. Detection and measuring of these kinds of mass phenomena require more attention to the development of accurate and comprehensive diagnostic criteria. As an example, consider the issue of domestic violence that is a social problem and can be expressed in many different terms such as physical, sexual, emotional, and economic abuses, with varying intensity and magnitude.[15] Moreover, since the concept of domestic violence is not uniformly defined in a similar way in different communities and cultures, it will be quite difficult to compare its occurrence across societies and cultures. Thus, conducting meaningful epidemiological studies across time and communities will require special effort and highly-focused attention to the details. Finally, regardless of how much effort is devoted to the development of comprehensive criteria for identification of cases of any mass phenomenon in a population, one can always be confident that the identified individuals do not represent all cases and not all identified individuals are, indeed, true cases. In other words, achievement of hundred percent accuracy is not possible. So, the basic objective of case selection for any epidemiological study is to increase the number of true cases and reduce the number of non-cases in the study population to the highest possible level.

The Population

The other principal element in epidemiology is the population at risk, i.e., the population in which disease or the mass phenomena of interest happens. The population is the denominator for most indicators, and knowledge of its specifics play a significant part in many types of epidemiological studies. Basic population descriptors like the total number, their sex, race and ethnicity, age structure, and places of residence is generally obtained from census and population surveys.

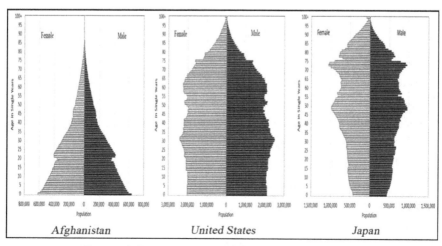

Figure 4. Major patterns of the population pyramid

The overall size and shape of population is mainly determined by birth, death, and to some extent, migration. The age and sex distribution of the population is generally presented in a graphic pattern known as a population pyramid and assumes different shapes in different communities. Figure 4 presents the population pyramids for Afghanistan, the United States, and Japan. The shape of the pyramid is determined by birth and death and is indicative of the overall condition of the community.

In areas with high fertility and birth, along with high mortality rates and lower expectancy of life, the lower part of the pyramid is very wide, and the upper part representing the older population is very narrow. These expanding populations generally face a problem with common infectious diseases of the younger people and issues with providing enough resources to support their rapidly-growing population. In areas with very low birth and low mortality rates, the population is called a contracting or an old population. Old populations have too few young individuals to maintain the cycles of reproduction and are saddled with a large number of elderly with high rates of chronic diseases that require extra care and attention. The old or contracting populations begin to shrink in their total number and eventually diminish as it is depleted by the elders death and not compensated with new births. Countries with stationary populations, where birth and death rates are in equilibrium, have a pyramid that looks more like a column. In these populations, health patterns and demographic profiles are stable, amenable to long-term planning, and void of serious population-related issues. The shape and size of a population is controlled by interaction of the three major components of births, deaths, and migration.

Birth

Births are the main source of population increase and is generally measured by various methods and presented as set indices in relation to the population. The crude birth rate is the ratio of the total number of births divided by the total population. Specific birth rates are limited to specific segments of the population like a particular ethnicity or the specific age of the mothers giving birth. The Hispanic birth rate is obtained by dividing the total number of Hispanic births by the Hispanic population in the area of interest. The number of births divided by the number of women in the childbearing ages of fifteen to forty-five years is called "fertility rate". It shows the average number of children per fertile women and directly determines the size and shape of the population. The optimum fertility rate that keeps the population stable is 2.1 children per fertile woman. In 2019, the global fertility rate was 2.4 with the maximum of 6.9 in Niger to a minimum of 0.92 in the Republic of Korea. Many countries in Africa had fertility rates of over 5, while many countries in North America and Europe had fertility rates of less than 2. The fertility rate for the United States is estimated at 1.84 in 2021.[16] Fertility rates that are higher or lower than the optimum, will result in changes in the population that is detrimental to its well-being. Increasing population size requires rapid expansion of resources, and decreasing population size risks the annihilation and disappearance of the society.

Death

Death or mortality is the natural way of reducing the population size. The proportion of all deaths to the total population is the crude "mortality rate". It can also be calculated within specific segments, including cause of death, sex, age groups, and so on. Over the years and with the advancement of medicine and public health, with overall improvement of physical and social environment, the mortality rate has universally decreased and in the United States, has tumbled down from 9.5 deaths per 100,000 population in 1950 to 8.8 per 100,000 in 2019. Apart from the overall numbers, the impact of mortality in a population can be best measured in terms of "life expectancy", that is the number of years a newborn child is expected to live. This indicator has globally improved over time and in the United States, has increased from 68.7 years in 1950 to 79.11 years in 2019. Life expectancy is a practical indicator of the advancement of health and improvement of the quality of life. The lowest and highest life expectancies in 2019 were recorded at 54.36 years in the Central African Republic and 85.29 years in Hong Kong.[17]

Migration

Migration is the permanent relocation of individuals, either internally within a given country or internationally between countries. Internal migration is generally from the rural to the urban areas or, rarely, from crowded cosmopolitan cities to suburbs and satellite communities. Migration has two aspects to it: "immigration" referring to those who arrive into the new community, and "emigration" referring to those who leave the community. It is controlled by the pull of the economic and social well-being of the target communities or the push of poverty, injustice, and poor living conditions of the home community. The effects of migration on a population are more qualitative than quantitative and tend to have significant impact on the socio-cultural characteristics of the communities involved rather than their size. Many host countries have immigration policies that are based on their specific needs. Migration is a major source for incorporating new talents, workforce, and traditions into the host countries, as well as the depletion of capable labor and educated individuals in the original countries that is generally known as "brain drain". The cultural impact and scientific contributions of the large ethnic immigrants in the United States is clearly visible in every

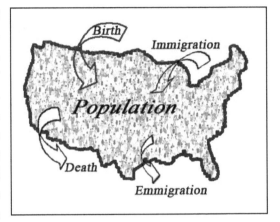

Figure 5. Factors impacting population

aspect of the life, from culinary and food industry to entertainment, advanced sciences, and technology.

In epidemiology, the "migrant studies" have offered valuable insights into the relative contributions of genetics and environment in disease occurrence. For example, Japanese migrating from Japan to the United States were noted to experience lower rates of stomach cancer and higher rates of colon cancer compared to those in Japan as they became acculturated to the U.S. environment.[18]

4 — THEORETICAL CONSTRUCT

The theoretical concept of disease production in epidemiology is based on the interaction of a triad composed of the host, i.e., people who are affected; the agent, i.e., the factor that causes the event; and the environment, i.e., place and time that influence and facilitate the occurrence of the event.

As noted in Figure 6, a disease (or any other phenomena) can only happen when all three entities are present and interact. The presence of the host and the agent or risk factors when there is no environment to provide contact and transmission between the two (HA) does not produce the disease. Numerous researchers work with killer microbes on daily basis in well-organized laboratories, and none of them are diagnosed with the disease, only because they are well protected and are not exposed to the agent. Similarly, combinations of host and environment (HE), where agent is absent, or environment and agent (EA), where host is absent, do not result in production of the disease or the event of interest.

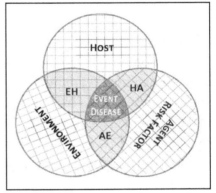

Figure 6. Epidemiology triad

A significant characteristic of the occurrence of mass events in large populations is their patterns of occurrence. A single case might be random and brought about by bad luck, but the occurrence of a disease or a phenomena in a large proportion of the population is not random and follows specific patterns that are influenced and at times determined by the characteristics of the host, the nature of the agent, and specifics of the

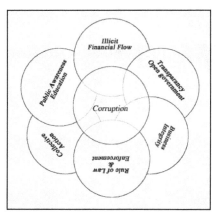

Figure 7. Multifactorial causation of Corruption

environment. For example, death is inevitable, and everybody eventually dies, but its pattern of occurrence is not the same for all people, in all places, and at all times. Men die more often and generally live a few years less than women. Death in young children is generally caused by infectious diseases, whereas in older people, it is more commonly associated with a chronic condition. Mortality patterns also differ by time and location. People in low-income countries are at higher risk of untimely death, drowning is more common in the summertime, and malnutrition is more frequent in low-income populations. These kinds of patterns are not unique to death and are present and detectable, with some variations, in all diseases and mass phenomena. One of the major tasks of epidemiologists is to detect these patterns and decipher the reasons for them. This will help with the development of proper methods for intervention and modification of the occurrence of the phenomena in the desired direction.

The Agent

Any factor that is needed for producing a diseases or an event is technically the cause of that event. There are three major types of causes for diseases in humans: biological, non-biological, and social. Biological causes are generally known as infectious agents and are responsible for diseases that can be transmitted from one host to other hosts. They can be transmitted directly from human to human like measles, viral hepatitis, and the common cold, or from animal to human, in which case they are called zoonoses. Many of recent epidemics that have originally been transmitted to humans from animals, like that of

SARS (severe acute respiratory syndrome) and COVID-19 from bats, MERS (Middle Eastern Respiratory Syndrome) from camels, and swine flu, are examples of zoonoses. Infectious agents can be transmitted among humans by means of the respiratory system, known as airborne transmission, and close contacts, like sexually-transmitted diseases. Other means of transmission include intermediary animal like insects, known as vector, and inanimate objects like clothing, utensils, and other personal belonging, known as fomites. Malaria and yellow fever are examples of vector transmission. Diphtheria, and most viral disease are transmitted by fomites. Organisms that are capable of producing infectious diseases are viruses, bacteria, fungi, and parasites are generally unicellular. Single-cell parasites are known as protozoa, and multicellular parasites known as metazoa. Some of the more frequent agent groups and diseases caused by them are listed in Table 1.

Table 1. A sample of diseases caused by biological agents

Biological Agent	Disease
Viruses	Measles, Hepatitis, Herpes, Influenza, COVID-19, etc.
Bacterium	Diphtheria, Whooping Cough, Tuberculosis, etc.
Fungi	Aspergillosis, Coccidioidomycosis, Candidiasis, etc.
Protozoa	Malaria, Toxoplasmosis, Chagas disease, etc.
Metazoa	Ascaridiosis, Schistosomiasis, tapeworms, flukes, etc.
Ectoparasites	Lyme disease, Plague, Scabies, Pediculosis, etc.

Non-biological agents include a large collection of hazardous chemical and physical elements. Hazardous chemicals, or poisons, can be produced in living organisms like nicotine in the tobacco plant, botulinum and tetanus toxins by bacteria, and aflatoxin by fungi and molds. Other groups of toxins include minerals and man-made elements like lead, mercury, pesticides, and various kinds of air pollutants. Physical agents include noise, vibration, radiation, and extreme temperatures. Examples of social agents include poverty, lack of education, and willingness or unwillingness to engage in risk-taking behavior.

Diseases produced by infectious agents generally follow a straightforward process with the agent being the sole cause of the disease, known as "necessary and sufficient". This characteristic can also apply to other disease agents like poisons, blunt force, or extreme temperature. For most non-infectious diseases and many social phenomena, however, causation is "multifactorial", which means there is a combination of various elements with different levels of impact. These are called "risk factors", which are contributory in the process of causation and while relevant to the process, are neither necessary for all cases nor sufficient by themselves. Multifactorial causation is observable in many chronic diseases and other mass phenomena. Figure 7 presents a simplified diagram of some of the risk factors involved in production of a multifactorial mass phenomena known as "corruption".

Inherent in the concept of multifactorial causation is involvement of different factors with different intensity in different communities. This situation provides multiple opportunities for intervention and at the same time can create major obstacles toward effective overall control and elimination.

The Host

The host refers to individuals who can be affected with the disease or the mass phenomenon of interest. Some of the characteristics of the host that are known to be associated with the occurrence or clustering of a disease or mass event are sex, age, race and ethnicity, occupation, education, marital status, socio-economic standing, and personal behaviors. Some of these factors like sex, age, and race are biologically determined and cannot be directly modified in relation to the occurrence of some diseases. Other characteristics such as education, occupation, income, and place of residence, and ethnic behavior are modifiable. Patterns of the occurrence of disease or mass events by host characteristics provide clues toward determining the causation and finding proper methods for implementation of the desired intervention. Most events occur differently in men and women, a difference that is measured in terms of the sex ratio, which is the proportion of male cases to female cases. At any given time, more males are born and more of them die, making the sex ratio for birth and death more than one. This was first noticed and reported by John Graunt in London in 1662[19] and still is a valid observation across the globe with some variation. In 2017, at the global level, 107 boys were borne for every 100 girls.[20] Similarly, the mortality sex ratio was 1.026 in the United States, 1.046 in Australia, and 1.203 in Egypt.[21] The net result is that women tend to live longer and have a longer life expectancy at birth.

Even though sex, age, and race/ethnicity cannot be directly modified, they can point to other areas of meaningful interventions. For example, breast cancer is significantly higher in women and a sex change operation cannot be considered as a practical means of prevention. However, one of the major approaches to combat this disease is routine mammography and screening women for early detection, and medical intervention is appropriate.

On the other hand, the higher frequency of major infectious diseases among the non-Hispanic black people is mostly because of their lack of access to proper medical care, which can be modified by provision of more equitable healthcare and preventive services.

A few of the other host characteristics that can influence the occurrence of disease or mass phenomenon are amenable to modification and change. These factors include education, occupation, access to support services, certain behaviors like smoking, and underlying diseases like blood pressure, and obesity that are generally acquired and can be targeted for modification and elimination.

Table 2 presents the association of suicide as a mass phenomenon by major host characteristics. It reveals that the overall age adjusted suicide rate is 13.64 per 100,000 in the United States and widely varies by sex, race, and age groups.[22]

Table 2. Suicide in the US by sex, race/ethnicity and age 2001-2015

Character	Classification	Deaths	Rates *	Percent
	Total	544,115	13.64	100.00
Sex	Female	115,770	5.68	21.28
	Male	428,345	22.34	78.72
Race/ Ethnicity	White (NH**)	455,995	16.71	83.80
	Black (NH)	31,152	6.35	5.73
	AIAN (NH)	6,033	18.37	1.11
	API (NH)	13,561	6.76	2.49
	Hispanic	37,374	6.75	6.87
Age specific by group(years)	10–14	4,264	1.36	0.78
	15–24	67,042	10.42	12.32
	25–34	83,813	13.69	15.40
	35–64	298,422	16.69	54.85
	≥65	90,574	15.19	16.65

* Rates per 100,000 are age-adjusted, 2000 US standard population, except for age groups. ** NH: Non-Hispanic. AIAN: American Indian/Alaskan Native, API: Asian Pacific Islander

As an another example, while coffee drinking is common in most parts of the world, it is also linked to various disease, including type 2 diabetes. In a study in Finland, the association of daily coffee drinking with a few of acquired characteristics was examined and is presented in Table 3.[23]

A few of the characteristics presented in this table, like smoking, education, and obesity, are acquired and can be manipulated for a desired outcome. This table also suggests the multifactorial nature of the determinants of coffee drinking.

The Environment

Table 3. Pattern of Coffee Consumption, Finland 2004

Cups of Coffee / day	<2	5-6	>10
Total Number	2,637	4,712	1,242
Mean age	49.2	48.9	46.1
Mean Education	9.9	9.0	8.5
% Tea Drinker	67.0	25.3	11.2
% Alcohol Drinker	50.4	46.7	50.5
% Current Smoker	11.5	21.4	50.6
% Obese	18.7	21.5	21.8
Education: total number of years completed			
Tea: one or more cup per day			
Smoking: one or more cigarette per day			
Obesity: Body Mass Index of 30 or more			

Environment is a general label given to all physical, biological, social, and chronological surroundings of the human host. The main functions of the environment are *maintenance* and *transmission* of the agents, causes, and risk factors of diseases, and it plays a significant role in the dynamics of epidemics.

Maintenance: The environment can function as a breeding ground for animals whose diseases can be transmitted to humans which are generally known as zoonoses and include major diseases like rabies, milk fever (brucellosis), and severe acute respiratory syndrome (SARS). It can also maintain favorable conditions for various arthropods like ticks, mosquitoes and other insects that transmit the agents of diseases like Lyme, malaria and yellow fever. It can function as a source of noxious materials like heavy metals and pesticides in the environs of manufacturing or agricultural areas. Conditions in some environments also enhance social breakdown, crowding, and deterioration of sanitation, or cause accumulation of hazardous material like pollution in the air and water that can maintain or create significant health risks to the population.-Sick people are also a source of maintaining and transferring infections. Individuals with clinical symptoms of a disease are easily identified and are isolated and treated. Some infected individuals, however, do not show any obvious sign or-symptoms and can transmit the agent of the disease while appearing healthy. These people are referred to as "carriers", and one of the famous cases of a carrier in the United States was Mary Mallon. She was a cook in the early 1900s in New York City and carried the typhoid bacteria with no clinical symptoms and caused typhoid epidemics anywhere she worked. She did not follow health recommendations and was apprehended by the New York Health department and spent the rest of her life in quarantine. [24]

Transmission: Transmission of agents or causes of disease is a major function of the environment and it assumes different modes for various agents and risk factors. Biological agents may be transmitted from infected individual or animals to healthy individuals in a few different ways, including: direct contact between the sick and healthy individuals, through contamination of inanimate objects like utensils or personal effects that are categorically labeled as "fomites"; airborne transmission that is mainly for respiratory diseases. Airborne transmission is done by means of droplets (larger mucoid exhaling particles when people cough) that travel a short distance of few feet, or by droplet nuclei (the dried out center of the droplet) that remain airborne for a long time and travel long distances. Covid-19 and its variants are a clear example of diseases that are transmitted by droplet nuclei and necessitate using mask, avoiding crowded areas and constant air circulation in closed places. Pneumonia, on the other hand is transmitted by droplet and social distancing and maintaining personal hygiene is a significant step in reducing its transfer. Other ways of transmission of infectious agent is with the help of vectors, that can be biological or mechanical. Biological vectors are insects like mosquitoes that transmit diseases like malaria and yellow fever, ticks in the case of Lyme disease, and house flies that can transmit various agents like those for the gastro intestinal diseases. Another common way of transfer is direct and physical contact between the sick and healthy individuals that is the main method for few viral diseases, including the sexually transmitted illnesses. Another significant way of transmission of infectious agents of gastro intestinal diseases is called "fecal-oral", and is widely occurring in areas with poor sanitation and low personal hygiene practices. The main medium in fecal-oral transmission is contaminated water. Many diseases like Cholera, typhoid, and diarrhea are transmitted through this route. A classic example is historical epidemic of cholera in London in 1849 that was traced to contaminated water from a certain pump on the Broad Street and was eventually eliminated following the closure of the pump by removal of the handle by John Snow.[25] Infectious agents can also be transmitted from animals to humans. Zoonotic diseases are usually common among people who live in proximity to animals. Rabies and swine flu are two major examples of this kind of transmission

Exposures to non-biological agents like poisons, physical trauma, accidents, electromagnetic radiation, etc., are mostly associated with places of residence and one's occupation. Living in agricultural areas increases the risk of exposure to harmful pesticides, while working in large industries increases the exposure to chemicals, extreme temperature, sound blasts and brute force. Living in poor neighborhoods, slums, and skid rows in large metropolitan areas increases the risk of exposure to violence, social unrest, illicit drug use, and similar hazardous situations.

Table 4. Major Types of Disease Transmission

Direct Contact	Human to human	Sexually transmitted Diseases
	Animal to human	Rabies
Environmental	Physical	Trauma
	Chemical	Poisoning
	Radiation	Cancer
Airborne	Droplet	Pneumonia
	Droplet Nuclei	Covid019
Waterborne	Bacterial	Diarrhea
	Parasite	Amebiasis
Vector borne	Mosquito, Tick	Malaria, Lyme Disease
	House Fly	Diarrhea

Regardless of it's role in disease transmission, the concept of the environment can be defined in three major ways: the natural, the administrative, and the social. Natural environment refers to areas that are distinguished by natural borders like rivers, mountains, lakes, deserts, and other topographical characteristics. The association of human diseases with natural environment has long been recognized and is mostly limited to zoonotic or other diseases transmitted by arthropod vectors. The classical example is the name malaria for the disease that literally means "bad air" in swampy areas. Other examples include the Aleppo sore (Cutaneous leishmaniasis), Malta fever (Brucellosis) and many of the viral diseases like Lassa fever, Rift valley hemorrhagic fever, and Ebola that are named after the geographical locations in which they were frequent or first identified.

Administrative environments are those described by political organizations and include international borders and national divisions of state, county, city, and zip code. The advantages of administrative classification include specific and measurable land area plus detailed information on the number, demographics, and other characteristics of residents, which is quite important in epidemiological assessments and investigations. Figure 8 presents the distribution of female breast cancer incidence in the United States by state.[26]

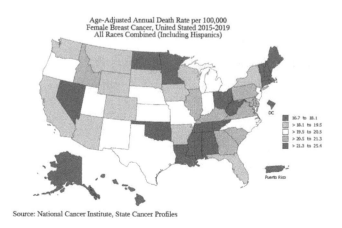

Figure 8. Breast Cancer Incidence in the United States 2013--2017

Finally, there is conceptual classification known as social environment that is based on a combination of a number of variables such as total population, dominant behavior of the residents, and social structure. A few examples of this classification include urban versus rural, metropolitan versus suburbs and satellite cities, as well as agricultural versus industrial communities. Studies have shown higher frequencies of injecting opioid drugs and syringe sharing along with increased frequency of hepatitis C in the inner cities compared to suburbs.[27] In the same context, the overall social and political environment of an area can have significant influence on patterns of disease and other mass phenomena.

Universal health care is a concept about access to basic resources for advancement and maintenance of population health and is highly dependent on the socio-political environment. It requires close collaboration among political leaders, health professionals, healthcare advocates, and policymakers.[28] In areas where such collaborations exist, most indicators of health are in good standing. This is clearly noticeable in comparison of the developed and still developing areas.

Time is another environmental variable that has a significant impact on chronological patterns of diseases and mass phenomena. Occurrence of mass events can increase, decrease, or remain constant over time. Rapid increases over a short period are indicative of epidemics that are generally sudden and are due to major and unexpected changes in the balance of the host, agent, and environment. Changes over long periods, years and decades, are known as the secular trend and represent the impacts of forces that control the increasing or decreasing dynamics of the disease. Secular trend is a good indicator of the effectiveness of interventions. Figure 7 presents the secular trends of cancer incidence and mortality in men and women in the United States from 1975 through 2020.[29] As noted in this graph, cancer mortality is declining in both sexes, which is a result of early diagnoses and proper treatment. Cancer incidence in men shows an abrupt rise in 1990s, which was caused by adoption of a blood test for screening (i.e., PSA) and detection of early stages of prostate cancer. This screening technique picked up a large number of medically-insignificant cellular abnormalities and was later abandoned, except in high-risk subsets of the population. While prostate cancer is a clinically important diagnosis in younger adults, many older males die with it, rather than of it.

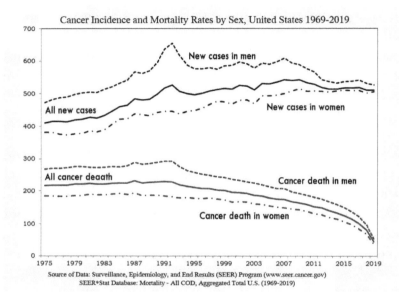

Figure 9. Cancer incidence and mortality rates by sex in the United States 1975--2017

5 — PATTERNS OF OCCURRENCE

Disease occurrence follows two major distribution patterns: general and specific. General distribution considers the patterns of the disease or event in the community, while specific distribution refers to the patterns of the disease in relation to the host, agent, and the environment.

General Distribution

Occurrence of disease or any mass phenomenon in a community can be described as sporadic, endemic, epidemic, pandemic, and controlled.

Sporadic is the situation when cases of disease are not common and do not follow a mass occurrence pattern. It happens occasionally under special circumstances when the host and the agent come into contact. An example is the Lyme disease or the bubonic plague; the agents of which are common in wildlife and can be transmitted by arthropod vectors to humans while visiting the wilderness. *Endemic* refers to a situation when the overall presence of the disease in community remains stable over time with no significant fluctuations. This means that an equilibrium is achieved among the three major components of host, agent, and

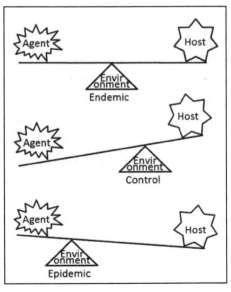

Figure 10. Balance model in disease distribution in a population.

the environment. In endemic situations, the number of new cases that are added and those deleted by death or treatment are almost equal. This results in a stable number of cases over time. Depending on the proportion of cases or death in the community, the endemic situation can vary from low to the very high level of hyperendemic. *Epidemic*, also known as *outbreaks*, refer to the situation when the balance of endemic condition is disturbed. This can happen by aggressive changes in the volume or character of the agent, significant decline in the defensive mechanisms of the host, massive changes in the environment in favor of the agent, or introduction of a new agent like that of the COVID-19. As a result, the number of new cases in the community exceeds the number that are generally removed by various means, resulting in the overall increase in the record of cases *beyond the expected* numbers, which is by definition an epidemic. Generally, epidemics have a defined time period and follow a pattern that can be visualized as a curve with three specific segments: the increasing segment when cases show up and rapidly propagate, the peak or the plateau which is usually of short duration, and a decreasing segment when the cases start to vanish. The curve shape of the epidemic is dependent on two important components: the number of susceptible individuals in the population and the extent of preventive interventions. With no intervention, the epidemic expands as long as susceptible individuals are available. Once the susceptible individuals in the community are exhausted or the transmission route is broken, the numbers of new cases begin to decline and the descending part of the epidemic curve starts. It is important to know that if the agent changes its nature, like Covid-19 virus that develops new variants, each variant acts as a new agent and produces new curves and peaks. Intervention activities include personal protection with vaccination and environmental modifications that are specifically designed based on the nature of the disease involved. Inherent in this definition is the fact that there are no magic numbers for identification of the epidemic. Any significant increase over the expected number is considered epidemic. Where no cases are expected, a couple of cases of disease may be considered epidemic. Similarly, where the occurrence of a number of cases are normally expected, significantly higher numbers are required before it can be considered an epidemic. Epidemic situations are generally limited to short time duration and specific geographical areas, such as epidemics of food poisoning in parties or occasional contamination of medical instruments in a healthcare setting. If the epidemic situation continues over long periods, they tend to change into an endemic level, like the epidemics of gun violence and misuse of the opioid medication that started as an epidemic but with over 45,000 gun related, and over 100,000 opioid related deaths in 2020, has currently reached the stage of hyperendemic among certain group of hosts in the United States.[30,31] If the epidemic situation expands over large areas and crosses the international borders, like the situation with COVID-19, then it becomes a *pandemic*. Currently, in the United States, major sexually-transmitted diseases are endemic, opioid death and COVID-19 are epidemic, measles is sporadic, and poliomyelitis is eliminated. Additionally, spread of COVID-19 over international boundaries to all countries across the globe is considered a pandemic.

The significance of an epidemic is that it represents an imbalance in the equilibrium among the triad of disease production and as such, is an unusual situation that requires unusual actions. Generally, epidemic control activities are preplanned, documented, and kept updated and ready for rapid deployment and use. Developing epidemic control plans during the progress of an epidemic is not a good idea. The COVID-19 pandemic clearly showed that those countries that had control measures on the books, because of their previous experience with SARS and were ready to implement them, had a much better experience in successfully controlling the epidemic. Epidemic preparedness is now a major component of proper health planning, which saves lives and protects the economy.[32]

Control of disease happens when the interaction among the elements of the triad is manipulated in favor of the host. This can happen by increasing host resistance, generally by immunization and physical protection, or by weakening and destroying the agent by proper medication, or by sanitation and modification of the environment to reduce or eliminate the transmission of agent or exposure to risk factors. These actions can be taken individually or in combination for more effective result.

Specific Distribution

The occurrence and spread of a disease or mass event, according to various characteristics of the agent, host, and the environment, is defined as the *specific distribution*. These characteristics are generally referred to as determinants that follow different patterns for different diseases or events in different communities.

Learning about the determinants of a disease or any adverse event is crucial in selection of appropriate intervention and control method. Opioid addiction and overdose has a set of determinants and follows a pattern that is significantly different from that of infectious diseases like influenza and measles. Opioid overdose death is a serious public health event and has a very strong behavioral component that is shaped by the socio-cultural structure of the society and the status of individuals in it. In the United States, 568,699 cases of opioid deaths have been recorded between the years of 1999 and 2015 that show 300% and 350% increase in 2015 over 1999 in men and women, respectively. The bulk of opioid overdose death is caused by Fentanyl and cocaine; has occurred in white men of twenty-five to fifty-four years of age; and in metropolitan communities of southern part of the United States. More details about the host, agent, and environmental determinants of the 46,802 drug overdose deaths reported in 2018 are presented in Table 5.[33]

Table 5. Determinants of opioid-related death in the United States, 2018

Determinant	Classification	Number	Percent
Sex	Male	32,078	68.54
	Female	14,724	31.46
Age group	0–14	65	0.14
	15–24	3,618	7.73
	25–34	12,839	27.44
	35–44	11,414	24.39
	45–54	9,565	20.44
	55–64	7,278	15.55
	65+	2,012	4.30
Race/Ethnicity *	NHW	35,363	75.99
	NHB	6,088	13.08
	Hispanic	4,370	9.39
	AIAN-NH	373	0.80
	API-NH	345	0.74
Community	Large Metropolitan	14,767	31.55
	Large Metro Fringe	13,476	28.79
	Medium Metro	10,328	22.07
	Small Metro	3,379	7.22
	Micro (Non-Metro)	3,162	6.76
	Non-Metro, Rural	1,690	3.61
Area	Northwest	12,467	26.64
	Midwest	11,268	24.08
	South	16,413	35.07
	West	6,654	14.22
Drug **	Prescription opioid	14,975	32.00
	Heroin	14,996	32.04
	Fentanyl	31,335	66.95
	Cocaine	14,666	31.34
	Methamphetamine	12,676	27.08
	Benzodiazepines	10,676	22.81
	Antidepressants	5,064	10.82

* NHW: Non-Hispanic White; NHB: Non-Hispanic Black; AIAN: American Indian Alaskan Native; API: Asian Pacific Islander
** Because of multiple drug use, these percentages do not add up

6 — DISEASE ACCOUNTING

Epidemiology is principally based on numeric data and the field of statistics is an integral part of it. Statistics may be defined as the science and art of collecting, managing, analyzing, and "making sense" of data by extracting and presenting the information embedded in numbers. The information extracted by statisticians is next translated to knowledge by epidemiologists and eventually results in action by various administrators.

Statistical methods cover all sorts of calculations and analyses from simple tabulation to elaborate multivariate regressions and advanced modeling. The field of statistics is instrumental in determining the true distribution of diseases, measuring the associations of various factors observed in such distributions, and evaluating the observed differences among various measures of the same entity. In fact, without statistics, many of the data-based conclusions cannot be achieved. Statistics

Data is all over the place and confusing! We need statisticians to make it neat as pin.

is necessary for the epidemiologist to maintain a reasonable grip on understanding data, developing intervention approaches, and evaluating their outcome. Nevertheless, statistics does not determine the importance or the relevance of the results to the objectives of the study. In a sense, statistics in epidemiology is similar to the microscope in pathology. Microscopes, from simple light-sourced to advanced electron-based, help with observing the tissue structure in detail with high accuracy, but the microscope does not make any decision about the value or importance of the observed changes in the pathology of the diseases. It is incumbent on the pathologists to put various facts together and arrive at a reasonable diagnosis.

Data in epidemiology is presented as counts and measures and to be useful, must be based on concise definition for accuracy and comprehensive coverage of all cases in the area of interest. Analysis of data that lack detection accuracy and is based on partial coverage will produce results that are biased and misleading. The value of the outcome of statistical analysis is directly dependent on the nature of the available data. Garbage data can only produce garbage results.

There are two basic approaches in statistical manipulation of the data: descriptive and analytical. Descriptive statistics covers the distribution of data according to major characteristics like sex, age, ethnicity, locality, time frames, and the simple associations among them. Analytical statistics, on the other hand, covers complex associations and correlations among various groups and elements of data.

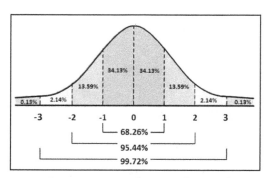

Figure 11. Distribution of normal population by variation segments

Most events or variables in a population follow the normal distribution of a bell-shaped curve as presented in figure 11. In a normal curve, the bulk of observations is centered in the middle of the curve, and the rest of it is equally distributed on either side. The distance between the smallest and the largest observations is the "range." Three other measures that describe the distribution are collectively called the measures of central tendency. These are the mean, median, and mode. Each one of them presents a different aspect of the data. The *mean* is the arithmetic average of the values of all observations. The *median* is the middlemost value that divides the total population of observations in two halves: above and below. The *mode* is the value with the highest frequency in the collected data. In a set of data that is perfectly normal or based on very large numbers, these measures are centered in the middle. In practice, however, these measures occupy different positions on the curve and their location on the curve describe the nature of the curve. If the mean is on the right side of the median, then the curve is negatively skewed, i.e. smaller values are more frequent. If the mean is on the left side of the median, then the curve is positively skewed, i.e. larger values are more frequent.

To determine the actual distribution of the data, a measure known as standard deviation (SD) is used. SD is a kind of index for measuring the range or the spread of data and is equal to the square root of the variance. The variance is the sum of the squared differences between the mean and individual observations. It can be used in multiples of 1, 2, and 3 (Figure 11) to estimate specific range of the normal curve, which is also known as the confidence interval (CI). When added to the mean, it provides the lower and upper

Should we brag about the mean height of our team, or state our median?

limits of the specified estimate for a given coverage. The mean plus and minus 1 SD covers 68.26% of the range of measurements in the collected data, the mean plus and minus 1.96 SD covers 95%, and 2.53 SD covers 99% of the area of the curve. Another measure that is generally used for evaluation and comparison of the results from various samples is the "standard error" (SE). SE is an estimate of the distance between the mean of a sample and the actual mean of the population from which the sample is obtained or the difference when comparing means from different samples. It has the same distribution as that of the SD on the normal curve. Mean plus and minus 1.96 SE identifies the two ends of an estimated span that will include the actual mean of the population, or the overlap of the means of two samples with 95% accuracy

Table 6 presents measures of central tendency based on the height of 17,650 American men. The mean height is 69.48 inches, and considering the SD of 15.1, it can confidently be estimated that 95% of observations are between the heights of 39.9 to 99.0 inches. Although the actual range was between 62 and 74, this implies that all observations considered in this example were within the range of normal height variation because none of them falls outside the estimated range. From another point of view and if one considers these observations based on a random sample of the general population, then it can be estimated, with 95% confidence, that the actual mean of the base population for this sample is between 61.9 and 77.1 inches.

Table 6. Measures of Central Tendency			
Range	Minimum	62	
	Maximum	74	
Mean		69.5	
Median		69	
Mode		74	
Variance		227.5	
Standard Deviation (SD):Övariance		15.1	
Standard Error (SE):ÖSD		3.9	
Confidence Interval (CI):		Upper	Lower
Mean ± 1,96 SD (95%CI)		99.0	39.9
Mean ± 1.96 SE (95% CI)		77.1	61.9

It is important to consider the implications of these indexes. If this was a group of soldiers in a camp, the SD would help the administrator organize and manage the particulars of the uniforms in such a way that there will be no shortages. If the group was a random sample of a larger population, then these numbers would help with understanding the height pattern in the population for research and comparative purposes. These measures play a significant role in making sense of data for description, comparison, or selection and monitoring of intervention approaches.

Measurements of diseases and other mass events in the community are either absolute or in relation to a population or another entity. Absolute numbers reveal the extent or the volume of the issue and are generally useful for administrative purposes. Once the number of patients is known, proper care facilities can be provided in sufficient numbers. Generally, numbers are larger in large communities, and that partially explains why care services are more abundant in big cities and metropolitan areas.

Another important aspect of statistical manipulation of data in epidemiology is to relate it with specific populations or "population at risk". Population at risk refers to the population from which cases of disease arose. In other words, it refers to a population that can generate or produce cases. As an example, the population of a general area like a county cannot be considered as the population at risk for cases developing in a city limit within the county. To be relevant, the denominator must be the population in the city. The two major types of relationships that exist between cases of disease and the population are the rate and the ratio. A rate is a measure of disease accounting that is based on the number of new cases which develop in a specific period divided by the population at risk of getting the disease in that period. This definition is called the incidence rate and presents the speed with which the disease progresses in the community. As an example, incidence of measles is calculated as the proportion of new cases over the population that are at risk of getting the disease, i.e., neither vaccinated nor having had the disease in the past, in a given period. Incidence is usually calculated for one year but can be for shorter periods as occurs with some epidemics and is expressed per hundred or some multiples of it for ease of understanding. The rate explains the speed of the development and progress of the disease or the event and is similar to the speedometer of a car.

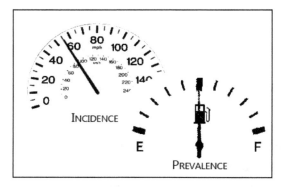

Figure 12. Incidence and Prevalence

In contrast, "prevalence", presents the momentary status of the disease in the community, and is a ratio. A ratio represents the relation between two entities. It can be the ratio of cases to the population at risk, professors to the students in a college, or physicians to population in a given community. Like a rate, it is generally expressed per hundred or multiple of it for better comprehension. The ratio of cases to the population, known as the prevalence. Like the gas gauge in the car, prevalence is an indicator of the magnitude of the disease.

Prevalence or the actual number of cases of a disease or event in a population is similar to the balance of a bank account. It is determined by the size and speed of the input, i.e., the number of new or incidence cases plus recurrences that are added, and the withdrawal, i.e., the number that is taken out either by treatment or death, Figure 13. Although clearly different, the combination of incidence and prevalence provides a clear picture of the extent of the disease and its speed of development in a population. At the beginning of epidemics, the incidence rate rapidly increases, and with effective control measures, it slows down.

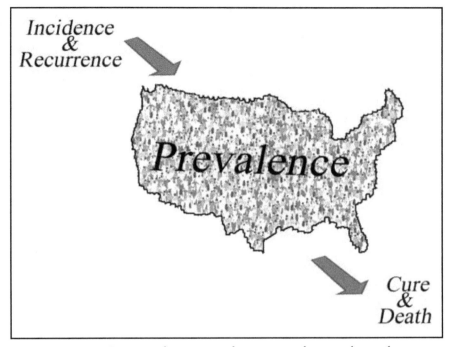

Figure 13. Conceptual association between incidence and prevalence

7 — TYPES OF MEASURES

The main measures in epidemiology are rate and ratio, and they can be calculated in three major types of: crude, specific, and adjusted. Crude rates represent the overall number of the disease or event divided by the total population. The crude birth rate in the United States in 2019 was 11.4 per 1,000 population. Crude measures are useful for finding the general distribution of the disease or the event of interest in populations. Around the globe, the crude birth rate fluctuates in relation to income from 10 per 1,000 in high-income countries to 34 per 1,000 in low-income countries.[34] Although better than simple numbers only, crude rates are still not sufficiently detailed or specific for comparing different populations, mainly because of the fact that different populations have different age, sex, and ethnic structure. Birth happens only in women and prostate cancer only in men, thus inclusion of men in the denominator when calculating crude birth rate or including women when calculating prostate rates is not appropriate. Apart from sex, most disease and mass events are highly associated with the age structure of the population. Since diseases and mass events do not occur similarly in all ages, comparison of the crude rates in populations with different age structures may result in false conclusions. Observed differences in crude rates in different communities may be more relevant to understanding the population structure than the nature and distribution of the disease.

One way to resolve this problem is to calculate specific rates. Calculation of specific rates is done by limiting the numerator (cases) and the denominator (population) to specific subgroups within the community. The fertility rate is an example of a specific rate where birth in a given year is related to the reproductively active population of women fifteen to forty-five years of age. Other specific rates can be calculated for different segments of the population for study or comparative purposes. However, use of specific rates for a given disease is not practical for comparison among populations because it produces a large number of results. Adjustment or standardization is an approach that allows us to combine multiple specific rates into a single measure that can be easily used for comparative purposes. Adjustment is generally done on important variables with a wide spectrum of distribution. Age structure is the most frequently-used variable for adjustment because it is significantly associated with the occurrence of most diseases and death. For sex, and race/ethnicity calculation and comparison of specific rates is not cumbersome. Age adjustment means imposing the same age distribution (from some standard or reference population) on different populations in the study. This equalizes the age distribution and produces a "constructed rate" that is similarly calculated across study populations and can be used for comparative purposes. This constructed statistic is referred to as the age adjusted rate and represents the rate under the assumption that the age distribution of the populations under study were similar. Although not actually exact, it is a good measure for comparing patterns across populations and times. A few standard populations are used, the most common of which are the US 2000 standard, US standard million, the World standard million, and the European Standard Population. The main difference in the standard populations is the proportions of various age groups in them. Data in Table 7 presents incidence of cancers of the lungs and bronchus in the United States in 2017 by sex, race, as age adjusted by different standard populations.[35]

Table 7. Comparison of numbers, crude, and adjusted lung cancer rates

Sex	Race	Cases	Population	Crude *	Age Adjusted	
					US **	World ***
Male	White	21,054	33,176,290	63.5	56.6	33.2
	Black	3,125	5,640,470	55.4	69.7	43.7
Female	White	20,887	33,425,800	62.5	47.0	28.8
	Black	2,712	6,078,022	44.6	43.9	28.3
* Crude Rate per 100,000, ** Rate per 100,000, age adjusted to the 2000 US Standard Population, *** Rate per 100,000, age adjusted to the World Standard Million (Segi)						

The differences noted in the age adjusted rates are because of the fact that the World Standard Million is much younger than the 2000 US Standard population, thus the lower rates for cancer incidence which is a disease of the older people.

Even in the United States, different standard populations have been used over time. The two major ones that have been used are the 1940 and 2000 standards. Table 8 presents the impact of these standard populations on the same data. As with all standard populations, the difference between the two US standard populations is in the age structure. The 1940 standard population has a larger proportion of children and young adults, which makes it a young standard. In the 2000 standard population, the proportion of individuals with advanced age is much larger and better represents the current demographics in the United States.

Table 8. Lung cancer mortality in California and New York, 2016

State	Sex	Ethnicity	Count	Population	Crude *	2000 **	1940 ***
California	Female	Not Hispanic	5,057	12,051,558	42.0	27.6	15.3
		Hispanic	563	7,322,068	7.7	11.4	5.9
	Male	Not Hispanic	6,025	19,103,036	45.5	35.9	19.6
		Hispanic	699	7,393,881	9.5	19.1	9.7
New York	Female	Not Hispanic	3,691	8,224,659	44.9	30.8	18.3
		Hispanic	201	1,820,353	11.0	12.8	7.0
	Male	Not Hispanic	3,856	7,702,625	50.0	42.0	23.9
		Hispanic	276	1,771,993	15.6	25.4	14.3

* per 100,000;
**: 2000 US Standard Population;
***: 1940 US Standard Population.
Source: https://wonder.cdc.gov/

Statistical and Clinical Significance

An important concept that is widely used and often misused is that of "significance". Technically, it refers to "statistical significance" and means that a sampled observation is out of the ordinary when compared to the general population or another sample. This observation is not a proof of the importance of the observed difference. It simply declares that with a certain level of error, the two samples are not the same. There are two types of significance: statistical and clinical.

Statistical significance is the declaration or the judgment that what has been found is of "out of the ordinary" in statistical analysis. This judgment is always limited (or even burdened) with error and is always qualified in a certain way. It must be accompanied with a statement on the amount of error that the researcher is willing to accept when making such judgments. Examples include the difference between the estimate of a parameter obtained in a study with its actual or true measure or between estimates obtained from different samples of the same general population. It is very rare to obtain the same estimates from different samples of the same population, and the researcher must make a decision to either call them the same or different. This decision is always made with some degree of uncertainty or error, which can be represented by the "confidence interval" (CI), which generally is set at 95%. Conversely, it may be noted as $p<0.05$. This means that a researcher accepts a probability of less than 5% error in making a judgment that the observed difference of the measures is out of the ordinary or not. This same concept applies to when different studies are compared. Considering data in Table 8, the statement of 5% suggests that if this study was performed on one hundred similar samples of this population, in five of them, the mean would be outside of the stated confidence interval, and that is the level of error that is accepted by the researcher. Occasionally, the level of accuracy is set at 99% ($p<0.01$). Thus, statistical significance actually represents the level of acceptable error on the judgment made by the researcher rather than importance of the observed measure. This is also true when two or more studies or populations are compared and the level of error that the researcher assumes in declaring that the measures being compared belong to the same or different populations. More recently, instead of declaring the decision by the researcher at a particular level, the upper and lower limits of the confidence intervals are reported to enable the reader to make a judgment on their own.

Statistical significance is influenced by the size of the population and the impacts of various kinds of biases or systematic errors in the measurements. Generally speaking, chances of obtaining a statistically significant result is higher in larger population samples because of the narrowing of the SD and SE as the sample size increases. Thus, where measures of large populations are concerned, the chances of obtaining statistical significance for very small differences in the measurements are high. Apart from the sample size, other kinds of biases can impact the measurements. A few examples of such biases are the difference in the structure and distribution of the study characteristics in populations, major differences in the nature and accuracy of the measurements based on clinical observations in comparison to laboratory measurements, and issues with the completeness and timeliness of reporting in various populations.

Clinical significance, on the other hand, refers to the actual and practical value and importance of the observed difference that may be labeled as statistically significant. While statistical significance is determined by statisticians, clinical or practical significance is determined by the clinicians or professionals who are directly involved with the issue and the basis of their decisions is more judgmental than mathematical. For example, in a comparative study of different treatments for cancer patients, two treatments for cancer of the pancreas were compared and were found to be statistically significantly different in increasing the survival at the error level of less than 5%. However, the actual difference in survival was only ten days, which has no clinical or practical value.[36]

Association and Causation

The next point to consider in dealing with statistical significance is the nature of the circumstances under which it is measured. In many cases, particularly in descriptive studies, statistical significance suggests that the simultaneous occurrence of two or more characteristics is meaningfully different from the expected patterns. This kind of statistical significance is known as an association. Associations are rather easy to find and at times are widely used and communicated, particularly in the media. They are often portrayed as some sort of evidence of causal relation and may exaggerate the importance of various kinds of interventions. While association is a necessary component of causation, not all associations are causal, and care must be taken not to confuse them. Causation is the situation when there is a temporal relation (i.e., time lag) between the two variables in a pattern of cause and effect. A cause or an "independent variable" predates the development of the effect or the "dependent variable". While associations are easily observed and recognized, causation is more complex, and its recognition requires significant planning and careful execution. The co-occurrence of smoking and gambling is an example of association since neither of them causes the other one. The association of smoking and lung cancer, on the other hand, is causal since smoking predates the diagnosis of lung cancer. Associations are necessary for causation but are not indicative of causation. Mixing of these concepts is the root of major misinterpretation of the statistical significance as evidence of causal relation.

8 RESEARCH IN EPIDEMIOLOGY

Any study or investigation of the nature and patterns of disease or other health issues and events in a population is, by definition, epidemiological in substance and practice. Generally, methods used for data collection and analysis is dependent on the subject matter and objective of the study, but the principles and the layout of the research are the same. Prior to the beginning of an investigation, a few major points that are collectively known as the study basics must be considered, reviewed, and adopted as the plan of investigation.

A. Study Question

Generally, questions to be pursued in epidemiological research are based on certain assumptions, which are based on information gained from prior investigations. Each study builds on the knowledge gained by previous studies and tend to expand the boundaries of understanding around a specific event and, in turn, will result in further investigations in the future. Current scientists are standing on the shoulders of past scientists and function as a pedestal for the future ones. Published reports in scientific journals are the main source of information for formulating new questions in conjunction with specialists in particular fields. Field specialists identify the general area of the question that needs clarification, and epidemiologists help with the design and conduct of the population-based study to provide relevant answers.

B. Study Population

The next step is to identify a population relevant to the study question. This can be limited to a particular group of inhabitants of various geographical entities, including county, state, and nation, or members of social or ethnic groups, such as Hispanics, Asian, and so on. Apart from its relevance and representativeness of a general population, selection of the study population is also influenced by availability and accessibility for participation and methods available for data collection.

C. Required Data

Data used in epidemiological investigations can be collected from different sources in different forms and types. It can be actively collected from the study subjects, or passively obtained from previously-recorded measurement kept in various databases. The two major sources of data are census (or total count) and a sample (or selected count). Regardless of the source, and to ensure the similarity of individual data points, a well-designed and concise instrument must be developed for data collection. Generally, this would be in the form of a questionnaire. Originally, questionnaires were printed on paper and manually completed. Currently, most questionnaires are electronic and computer-based. Electronic questionnaires are highly preferred because they are less prone to recording errors and are easier to manipulate and process. Hardcopy questionnaires are highly time-consuming and labor-intensive for verification and data entry and subsequent filing. Virtual questionnaires, on the other hand, are much easier to handle and less prone to data entry errors. For example, with the use of proper algorithms, most of the recording errors can be prevented, many of the calculations such as determining the age by entering the date of birth can be completed instantly. Gender-related questions will be properly handled, and the sequence of questions can be adjusted while in progress. A number of software programs have been developed to facilitate the construction of the questionnaire, help with data collection, and assist with analysis of the data and are currently in use by many researchers. Regardless of the type of questionnaire, care must be taken to avoid bias in developing the questions and collecting responses. Bias and/or intentional error occurs when the question format or the way in which a question is asked directs the subject toward a specific response that is predetermined or favored by the researcher. What do you think about smoking? is a neutral question that does not direct the respondent to a specific response. Do you agree that smoking is hazardous to your health? is a question that tends to direct the respondent toward a very specific response.

The sources of Data

"In order to accomplish the types of studies described above, whether the work be descriptive or analytical in nature, the epidemiologist must rely on various sources of data. Few of these sources are total counts of events in the population, of which Census of the population conducted every ten years by the U.S. Bureau of Census is an example. More often however, the epidemiologist must collect the required data from individuals usually by administration of questionnaires or interviews. This step requires a process known as sampling."

The Total Count

Total count refers to collection of data for all cases of the disease or the mass event in the community. This is similar to a census of the population and is cumbersome, time-consuming, and

Table 9. Fatal injuries and suicide in young adults in the United States, 2015

NVSS *	Deaths		
	Number	%	Rate **
Intent			
Unintentional	7963	61.4	9.7
Homicide‡	2544	19.6	3.1
Suicide	2470	19.0	3.0
Sex			
Male	8951	69.0	21.3
Female	4026	31.0	10.0
Age Group			
<1	1554	12.0	39.1
1–4	1604	12.4	10.1
5–9	895	6.9	4.4
10–14	1330	10.2	6.4
15–19	7594	58.5	36.0
Total	12,977	100.0	16.2

* National Vital Statistics System
** Rates are per 100,000 population
‡ Includes assaults, legal intervention and terrorism, but excludes operations of war.

expensive. The most well-known examples of total count are related to events like birth, marriage, divorce, and death that are legally reportable. Some specific infections that are less common, but important, and few chronic diseases and events of special interest are also legally reportable. In the United States, the Centers for Disease Control and Prevention (CDC) and the National Vital Statistics System (NVSS) [37] determines which diseases and vital events are reportable. Additionally, a few other organizations collect data on the occurrence of other kinds of events and maintain total counts related to health and welfare issues. The main reason for this approach is to monitor the dynamics of the disease in the community and to detect "icebergs."

As explained earlier, this phenomenon in epidemiology refers to the observation of a few cases of a particular reportable disease that is out of the ordinary. Like an iceberg in the ocean, it is indicative of a larger number of cases in the community. As mentioned earlier, detection of few cases of Kaposi sarcoma and pneumocystis pneumonia, in Los Angeles in 1981, was the iceberg that was noticed by the CDC epidemiologists and upon further field work, resulted in identification of AIDS and HIV infection. Apart from iceberg detection, the main objective of a total count of data is to determine the overall temporal and chronological patterns of the occurrence of disease and its probable deviations from the expected normal. Table 9 presents the distribution of 12,977 fatal injuries in population of under nineteen years of age in the United States in 2015.[38] As good and comprehensive as the total count sounds, it is not easily available or even possible in all instances. The anecdote or analogy is that to find out the total number of substandard products in a cargo container, one must examine all the items in the container! Moreover, if examination requires the destruction or modification of the item, a total count is not practical or advisable. The solution to this dilemma is sampling.

Sampling

Sampling is a highly-complex undertaking involving many steps generally designed in collaboration between statisticians and epidemiologists. Technically, sampling is the procedure of selecting a small, manageable, and representative segment of the population under study. The main steps in sampling include determination of the size of the sample, the reference population, and the process of selecting participants. The cardinal issues in sampling include two principles: (a) that it must be representative of the reference population, and (b) it must be random in nature. Samples that are not truly representative of the reference population or are selected in a non-random process are considered biased and produce questionable and erroneous results, that cannot be related to any specific population.

To be representative, the sample should mimic the general constructs of the reference population and its distribution with regard to major characteristics, such as sex, age, residential area, race, and ethnicity. Moreover, it must be selected in a random manner. Random selection means that each individual in the reference population has an equal and predetermined chance of selection into the sample and that this chance (or probability) is more than zero. Samples that do not reflect the overall constructs of the representative population or are selected in a nonrandom fashion are considered biased.

Sample Size

Determination of the sample size requires strong statistical knowledge and is generally achieved in collaboration with a statistician and is based on information about the actual or perceived frequency of the intended measurement, the desired level of accuracy, and an acceptable margin of error for the outcome measure. Suppose one wants to determine the actual proportion of smokers in a community of 10,000 people where the overall frequency of smokers is estimated to be about 10%.

A researcher wants to measure this proportion with a margin of error of 5% and accuracy of 95%. The margin of error refers to the level of fluctuation of the estimate from the actual measure. A margin of error of 5% means that the measure obtained by sampling falls within plus or minus 5% of the actual proportion in the community. If the actual proportion of smoking in the community is 10%, then the sample result that falls between 5% and 15% has an accuracy of 95%. This level of accuracy means that if this study is repeated one hundred times, in ninety-five of them, the results fall within the stated margin of error. Special formulas are developed that determines the sample size based on these parameters, and any change in them will change the sample size.

Sampling Frame

The sampling frame is the total or defined part of the population of interest, from which sample cases are directly drawn. A sampling frame is selected based on ease of access and unbiased reflection of the major characteristics of the reference population. As an example, consider selecting a sample of students at a university. Various sampling frames, such as the student ledger at the dean's office or students who pass through the gates on a particular day, or students who are present in a particular class, are possibilities. Each of these frames has its own advantages and shortcomings. The issue with the ledger at the dean's office is that once the selection is made, getting access to the selected students might not be possible for a variety of reasons. The passage of students through the gate is influenced by a number of different conditions that make the sample not representative of the general student population, and a class list tends to severely limit the application of the sample result to the general student population. These and other concerns about sampling frames should be weighed and evaluated by the researcher prior to the study.

Sample Selection

Once the sampling frame is selected, individual sample participants must be selected. There are many different methods for actual selection of individuals into the sample.[39] *Simple random sampling* procedure is applicable when the population in the sampling frame is manageable and can be enumerated. Once the population is counted and numbered, members can be selected similar to the lottery by using random numbers. *Systematic sampling* refers to selecting individuals in a regular sequence from the sampling frame. For systematic sampling of patients visiting a clinic, the first individual is selected by a random number between 1 and 10, and then it is repeated for every following sequence of ten patients. For example, if the first randomly-identified individual is patient number 4, then individuals numbered 14, 24, 34, 44, and so on who come to the clinic are selected into the sample 'til the predetermined sample size is achieved. *Stratified sampling* is considered when the reference population is composed of unequal groups of male and female, different employment status groups, age groups, and racial distribution. Under these kinds of circumstances, the proportion of each group in the reference population is identified and applied to the sample. If the reference population is composed of 40% female, then 40% of the sample must also be female. This concept applies to all other stratifications in the reference population. *Cluster sampling* is generally used for large populations like residents of a city. In this approach, the total population is divided into clusters of roughly-equal size and similar character. Next, a few of the clusters are randomly selected, and within each cluster, a predetermined number of individuals will be randomly selected and included in the sample. A good example of this approach is the child immunization survey developed by the World Health Organization that is based on 210 children selected from 30 clusters in a village or city. *Convenience sampling*, *snowball sampling*, and *quota sampling* are examples of non-random (non-probability) methods that are based on availability and willingness of individuals in the sampling frame for participation in the study. In convenience sampling, members of the sample are selected based on their agreement to participate in the study.

In snowball sampling, once the initial members are identified, they refer their friends and relatives to be included in the sample. Quota sampling is similar to stratified sampling with the exception that the selection is based on convenience rather than a random protocol. Non-random sampling is useful for the study of special issues like illicit drug use, homelessness, illegal immigration, and so on. Results from non-random samples are generally open to question because of the self-selection of respondents, their representativeness of the reference population, and the accuracy of the result as a valid estimate of the measure under study in the reference population. Some of these problems, however, can be resolved with advanced and elaborate statistical approaches.[40]

9 DATA HANDLING

Data Analysis

The next step following data collection is cleaning the data and data analysis. Cleaning aims to detect and delete illogical entries, duplicates, obvious factual errors, such as recording wrong sex for cases, or other data entry errors that will need to be double-checked and clarified. Data that are obviously out of normal and expected range need to be re-examined and handled as outliers. Once this step is completed, data is ready for analysis. Data analysis can extend from simple tabulation to highly-complex statistical modeling. Presently, a large number of computer software programs are easily available for data analysis (e.g., Stata, IBM/SPSS, R, SAS, etc.). It is very easy to use these software programs and calculate rates, ratios, or other parameters. The important point is to ensure the underlying concepts, the exact question to be answered, and a legitimate need for a particular type of analysis, are clearly understood and discussed prior to running the analysis. This step must be carried out in collaboration with the statisticians.

Data Presentation

Communicating the results of data analysis can be done in text, tables, graphics or a combination of these modalities. A good visual representation, such as a well-constructed table or accurately-designed graph, is highly valuable and can convey significant information in an easy and understandable manner. A good and appropriate graphic presentation is worth more than thousand words in conveying information. Throughout this document, examples of tables, figures, and graphics are included to help with explaining complex concepts. Some of the more frequently-used graphics for data presentation are shown in Figure 14.

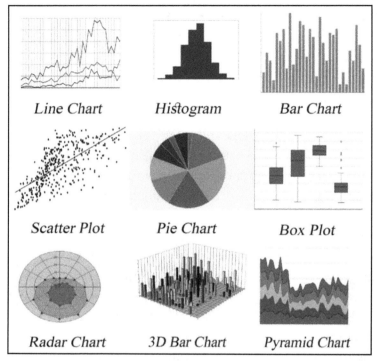

Figure 14. Example of common data presentation graphics

10 — TYPES OF EPIDEMIOLOGICAL STUDIES

Epidemiological studies can be described based on their temporal direction and objectives. Directionally, they can be limited to a given point in time or can cover periods of forward-looking or backward-looking in nature. Objectively, they can be descriptive, analytical, or experimental in nature, Table 10.

Table 10. General layout of epidemiological studies

Direction Classification	Cross Sectional	Backward- Looking	Forward- Looking	Cohort
Descriptive	✓	-	-	✓
Analytical	✓	✓	✓	✓
Experimental	-	✓	✓	✓

Study Direction

Study direction is the manner in which data are collected. Cross-sectional studies refers to collecting relevant data at a given point in time, which could be as short as a particular day or a specific period, like a year. Backward-looking or retrospective means looking back in time and collecting data on past events. Retrospective studies are based on either documentation of past events or memories of participants, which can introduce some problem with completeness and accuracy of the availability of data. Forward-looking or prospective means looking forward to collect data in the future as they unfold or develop. Prospective studies tend to be more accurate and can be highly-detailed but are more time-consuming, difficult and complex to carry out, and very costly. The term "cohort" refers to the concept of monitoring a large group of selected individuals over a long period of multiple years. Individuals selected into a cohort are identified based on their belonging to the group of interest identified by the organizers of the cohort. This is slightly different from the original definition of the term cohort which was applied to a group of individuals participating in prospective studies with different exposures to the independent (exposure) variable or the suggested cause and followed up to the occurrence of the dependent variable (usually disease or death) or expected effect. The main advantage of this concept is that all the collected data are diligently verified at the time of collection and properly classified and stored. Thus, biases related to the accuracy of the recall and incomplete collection that is a significant source of error in retrospective studies is diminished. Current use of the term "cohort" refers to a group of people with a common experience who are selected and followed for a long time to monitor the occurrence of various events (e.g., disease or death) in them. A few of the common experiences are birth in a particular year (birth cohort), participation in a particular event like military action in a foreign land, or having been diagnosed with a certain disease like breast cancer. Data collected on these populations can be used for different types of studies in all the three directions mentioned above. At the present, many cohorts have been developed around the globe to study all aspects of health, disease development, and social events. One of the most famous cohorts in the United States is the Framingham Heart Study. This cohort was initiated in Massachusetts in 1948 for the study of heart disease in the middle-aged and elderly population. Since then it has resulted in over 3,300 scientific papers and major discoveries concerning diseases of the heart.[41] In recent times and with the ease of electronic data collection, access to large databanks, powerful computers, and sophisticated software, it is possible to establish virtual cohorts and conduct large number of analyses with relative ease and accuracy. While the expansion of the electronic databases and sophistication of electronic data handling is very helpful in advancement of population health and prosperity, it can also increase the possibilities for misuse. A significant concern in this area is the "backward" analysis, which means analysis without specific question. Under this approach, a number of analyses using pre-developed programs are conducted, and once an interesting result is obtained, relevant questions for which the obtained results can be accepted as an answer are developed. In other words, the answer comes first and the question later. This is called data dredging, a scientific dishonesty and cheating, and is not accepted as legitimate research.

Study Classification

This classification covers the objectives of epidemiological studies and can be defined in three major groups of descriptive, analytical, and experimental.

Descriptive Studies

The main objective of this class of studies is to identify the characteristics of the disease or issue of interest as a mass event and to describe patterns of its occurrence in relation to major elements of host, agent, environment, and time. These studies are performed at the beginning of any meaningful epidemiological activity, provide a clear and concise picture of the issue, its distribution patterns, and its relative magnitude. Descriptive studies are generally cross sectional in nature and are limited to a specific point or period. They provide reliable evidence of associations that can be used as a basis for the development of analytical studies. A good example of this kind of result is presented in Table 4, from a cross-sectional study of death caused by drug overdose in the United States in 2018. It provides a clear picture of the patterns of drug overdose in terms of who it affected, where in the nation it is more common, and what kind of drugs are involved.

Regular descriptive studies are generally limited to a specific time, like one year, but can also cover longer periods of multiple years and can provide valuable information on chronological or temporal distribution of the issue of interest. Chronological presentation of certain indicators, like mortality rates, birth rates, and cause of death ratios, provides information that reveals the dynamic of the event over time, which is known as the secular trend. Figure 7 presents the temporal distribution of the incidence and mortality for major cancers by sex over forty-two years in the United States. This temporal distribution or secular trend reveals dynamics of these cancers in recent years.

Analytical Studies

The objective of this category of epidemiological studies is to establish a causal relationship and is generally based on statistically-significant associations identified in descriptive studies. Analytical studies by necessity must start with a hypothesis, which is the central point in its design and conduct. Briefly, a hypothesis in epidemiology is a statement of an assumed direct and positive relation between two events that are identified as the cause and the effect. The cause or exposure is labeled as the independent variable, while the effect or disease is labeled dependent variable. Although the aim of analytical studies is to confirm the hypothesis and causation, from the philosophical point of view, this cannot be done directly. The reason is that any proof of actualization of a hypothesized event in the past cannot logically prove its occurrence in the future regardless of how many times it has happened. Actually, if the experience confirms the validity of the stated hypothesis concerning the causal association, then there would be no need for further testing of it. Causation is generally hypothesized based on observed associations that are yet to be proven as causal. The solution suggested for this issue is to develop a "null hypothesis" that is the opposite of the original hypothesis and is based on the assumption that the stated causal association is false and does not hold. This approach necessitates that hypothesis in epidemiology context be "refutable". Hypotheses that assume causal association for which null hypothesis cannot be developed, i.e. are not refutable, cannot be scientifically tested. The concept of refutability is well explained in the classical example of the hypothetical statement that "all swans are white," the null of which is that "not all swans are white." Obviously, one cannot examine all swans and is limited to the observed cases. Thus, no matter how many white swans are encountered, there will be no assurance that the next swan will be white. However, as soon as a black swan is found, the hypothesis that all swans are white stands as false. Similarly, the hypothesis that "vaccination prevents disease" can only be tested when the null hypothesis that "vaccination does not prevent disease" is tested and refuted. In other words, testing hypothesis is based on the actualization of the null hypothesis, rather than validation of the hypothesis. This is similar to the concept of guilt in legal proceedings. As long as the guilt is not proven, the accused is considered not guilty. Analytical studies can be carried out in three different time frames, Figure 12.

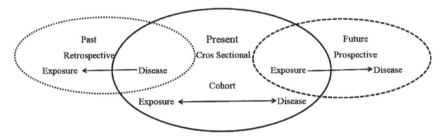

Figure 12. Schematic direction of different types of analytical studies.

Retrospective or "looking back" studies start with the affected (i.e., diseased) individuals and collect the necessary data on the presence of the hypothesized cause for events that may have happened in their past. Another name for this kind of study is "case-control". In testing the causal association between smoking and lung cancer, the history of smoking in a group of individuals with lung cancer (cases) is compared with the similar history in a group of individuals without lung cancer (control). Next, information about past smoking habits in the two groups is collected and analyzed. The causal relation between smoking and lung cancer will be accepted when the frequency of smoking is found to be significantly lower in the non-lung cancer controls. The major concern with this approach is the availability and access to accurate and complete recorded data for past events or accurate memory recall. Retrospective studies can be conducted over short periods, are not generally expensive, and are relatively easy to manage. Lack of access to reliable sources for complete and accurate data in the past can be a major hurdle for these kinds of studies.

Prospective or forward-looking studies start with the exposure to the cause or risk factor and then follow individuals until the occurrence of the effect. Participants in these kind of studies are technically known as the cohort. It is composed of a group of very similar individuals in major characteristics, who are then divided into two groups of those who are exposed to "cause or risk factor" and those who are not. The two groups are followed up and the occurrence of disease (the "effect") is determined in them. To test the hypothesis on the causal association between smoking and lung cancer, the two selected groups are very similar, except that one group is the smoking group and the other one is not. Prospective studies are highly accurate because the researcher collects and verifies data as it happens. Prospective studies require sufficient time for the cause to produce the effect, which, in the case of smoking and lung cancer, may be extended into years or even decades. Because of this, they are time-consuming and require a well-designed organization with significant human resources for follow-up, data collection, and handling, which makes them very complex and expensive to conduct.

Cross-sectional is a kind of study that is based on a large group of individuals who are observed at one point in time during which the presence of various risk factors as well as the resulting outcome is measured. Although cross-sectional studies were originally developed as descriptive, within the frame of large cohorts, it can be done in analytic context. Cross sectional studies describe the current status and by simultaneously measuring of the exposure and the effect, and they provide a good source for statistical association. For example, in California, there are fifty-eight counties, and on any single day, air pollution levels may be measured at the county level. Simultaneously, asthma attacks, as measured by emergency department encounters, may also be measured, and the correlation between the two variables may be calculated. This, however, cannot be accepted as a support for causal association, unless the collected data can clearly show that the air pollution actually happened prior to the asthma attacks and counties with no air pollution had far fewer asthma cases. In recent years, the concept of a cohort study in large populations has been expanded, from a specific research strategy for testing a single hypothesis to the study of the totality of one major issue or disease, as noted above for the Framingham Heart Study. Cohort populations are generally created around one major health issue, which have a complex causation with the objective of collecting all the data relevant to the selected topic. In this approach, because of the availability of accurate data on exposure to various independent factors or the cause and development of the dependent factor or the effect within the proper chronological order, analytical studies in both retrospective and prospective direction can be done in a short time with a very high accuracy. A major issue with longitudinal population cohorts is the collection, management, and storing of extensive and accurate data in special databanks and making them easily available to researchers. Although the development and maintenance of cohorts are expensive both in terms of manpower and resources, over time they provide a very reliable source of data for epidemiological studies. In recent times and with the advent of electronic capabilities, a large number of cohort studies have been developed across the globe and have extensively been used for studies in all fields of public health and other mass events, including political interventions.[42]

Relative Risk

Relative risk is a measure of association and one of the main indicators of altered risk in analytical and experimental studies. It measures and explains the ratio of the outcomes of interest in groups with different levels of exposure to a same risk, i.e., the ratio of lung cancer in people with different smoking habits, or smokers versus non-smokers. In analytical studies, relative risk is the ratio of the effect in different exposure groups under study. In experimental studies like clinical trials, the relative risk is the ratio of outcome in the test group compared to that of the placebo group.

In the following table, adapted from the COVID-19 vaccine trial with 43,448 participants, the relative risk is the ratio of diseased cases among the vaccinated group (A/A+B) over the ratio of diseased cases among the non-vaccinated (C/C+D). In this trial, only 0.037% of vaccinated individuals became ill, while of the non-vaccinated individuals 0.746% became ill. The relative risk (0.037/0.746) is 0.048%, which means that the vaccine used in this study had 95% effectiveness (1.000-0.048=0.952)in preventing the disease.[43]

Vaccine	Disease Yes	Disease No	Total
Yes	A	B	A+B
	8	21,712	21,720
No	C	D	C+D
	162	21,566	21,728

Experimental Studies

The objective of experimental studies is to evaluate the impact of selected interventions on the disease or mass phenomena in a population. Intervention in this context is an act that is expected to produce the desired outcome. Experimental studies can be considered an extension of prospective analytical studies with the difference that the nature of the intervention and the study environment are selected and controlled by the researcher. Experimental studies covering disease prevention or treatment are generally known as randomized controlled trials (RCT). RCTs have three important characteristics: a) are controlled because they include group(s) that receive the intervention (known as the experimental group), and similar group(s) that receives the placebo and are known as control group(s). Control groups are similar to the experimental group in all major characteristics, except for the intervention. The placebo is an action (or drug) very similar to the intervention in appearance that is used in the experimental group, yet has no active ingredient. It is an inert intervention and is used to prevent observation biases and to make sure that both groups are monitored and observed under similar conditions; b) are randomized because division of the participants into the intervention and control groups is randomly determined; and c)are considered a trial because they are conducted under the complete control of researchers.

The main outcome of interest in experimental studies is the measure of the effectiveness of interventions in modification of the outcome, which, in the case of disease, is prevention or treatment. Vaccination, new drugs or treatment regiments, environmental sanitation, health education, and enforcement of legal regulations are few examples of interventions used for modification of the occurrence of diseases or other mass phenomena of interest.

Phases of Randomized Clinical trials

Major concerns in experimental studies revolve around the safety of the intervention, informed consent of the participants, and accurate measurement of the effectiveness of the intervention. Safety is a common reference to the intervention in relation to its possible harmful impacts. Safety is technically determined by the occurrence of harmful effects on the recipient and is basically measured in terms of clinical events like fever, pain, etc. Informed consent is to ensure that participants are aware of the specifics of the study and are participating voluntarily. Development and testing the effectiveness of therapeutic interventions or vaccines is a complex activity and passes through specific stages or phases. These stages are proof of principle (phase 0), safety (phase I), effectiveness (phase II), efficacy (phase III), and post-marketing surveillance (phase IV).

Table 11. Stages of Clinical trials

Phase 0	Phase I	Phase II	Phase III	Phase IV
Proof of Concept Laboratory Work	Safety Test Medical Observation	Effectiveness Controlled Testing	Efficacy Randomized Controlled Trial	Delayed Side Effects Post Marketing
Mainly Laboratory work	Small Numbers under 50	Large Number 50 to 300	Large Numbers over 1000	General Population over 100,000

The proof of principles phase covers studies that provide some evidence of the impact of intervention on the issue in the desired direction. This step is generally conducted at the laboratory level and on test animals. It is commonly based on prior studies and observations by scientists in the field. The proof of principle is phase 0, and if it produces good and acceptable results supported by scientific evidence, then the study will move to the next step, which is the proof of safety phase. Proof of safety is Phase I, and its objective is to measure the side effects and discomfort experienced by participants after receiving the intervention in practical dosage. This phase is done on a small number of volunteers who will be under strict and close observation following the application of the intervention. Once the safety of the intervention is confirmed, the study can move to Phase II or proof of effectiveness, which is generally performed on lager number of volunteers under controlled condition. At this stage, the volunteers are divided into the experimental group who receive the intervention and the control group who receive the placebo. Individuals participating in this phase of study are either sick individuals who participate for testing new therapeutics or healthy volunteers to show the impact of preventive measures like vaccination. In this stage, the reaction of individuals in the two groups of intervention and placebo is measured and compared. Upon successful completion of this step, the final step or Phase III begins, which is to measure and evaluate the efficacy of the intervention in a large population of volunteers who are also divided into the intervention and placebo groups and are followed under natural conditions. Effectiveness and efficacy measure the same thing with the difference that efficacy is measured in a small group of subjects under controlled environment, while effectiveness is measured in a much larger population under natural uncontrolled situation. Under special circumstances, like that of the trial for COVID-19 vaccine, it is possible to combine Phase II and Phase III to expedite the development of the final vaccine. Upon obtaining acceptable results from Phase III, the intervention will be offered to the general population. However, recipients will still be closely monitored for the occurrence of rare side effects that could not be picked up in earlier phases. This is called Phase IV or "post-marketing monitoring", and the basic concept for it is the very large number of people who receive the intervention in the public and provide the opportunity to observe rather rare side effects. It consists of regular monitoring of those who receive the intervention, which is usually some sort of therapeutic agent. A well-known example for the need of this kind of monitoring is the thalidomide effect. This drug was first introduced in Germany in 1958 as an over-the-counter sedative, hypnotic, and remedy for motion sickness. Many pregnant women used it across the globe, including Canada, United Kingdom, and the United States. By 1962, a large number of children with various congenital malformations were born that was directly related to the use of this medication by their mothers during early months of pregnancy. This kind of monitoring is generally performed by the official organizations like the Food and Drug Administration (FDA) that, in 1961, identified thalidomide as the cause of congenital malformations.[44] A more recent example of the Phase IV study is detection of blood clots with a frequency of one in a million recipients of COVID-19 vaccine that could only be detected upon widespread dissemination of the vaccine in the general population.[45]

Placebo Control

Use of placebo in the control group is necessary for reducing the impacts of various observational biases that may enter the study. The technical jargon for administering a placebo along with the active ingredient is called "blinding", because it is done without the knowledge of study participants or investigators. Blinding can be imposed at various levels. To ensure that the knowledge of group assignment will not influence the behavior of the participants, the nature of the assignment group is withheld from the participants (single blind). Next, the group assignment is withheld from the observers to ensure that this knowledge does not influence their measuring of the outcome (double blind). Finally, to ensure unbiased analysis, the knowledge of group assignment is also withheld from the statisticians who performs the data analysis (triple blind). Upon completion of the trial and obtaining the final result, the blinding will be removed, and if the intervention has been successful, all the controls will also receive the drug or medication under study.

Informed Consent

Informed consent (IC) is a process developed to protect study participants from any harm by ensuring that they fully understand the experiment in which they are to enroll. This is generally done by providing a written explanatory document that is signed by the participants and the principal investigator of the trial. It is a legal and binding document. It confirms that the participant in the study is well informed of the nature, objectives, and possible side effects of the interventions planned in the study and has voluntarily and willingly agreed to participate in the study. This document must provide detailed information about the experiment and explicit contact information about the researchers conducting it. It must also clearly state that the participants will be getting either the actual intervention or the placebo on a random basis without their specific knowledge of it and that they can leave the study at anytime for any reason. This document must also include detailed information about the grievance procedure and methods for filing a complaint. Assurance must be obtained that participants have completely understood the meaning of the statements in the consent form and agree with the intentions of the study. Unless this informed consent is completely executed for all volunteers, the study cannot proceed.

Clinical Trial, Trial and Error

The phrase "trial and error" is sometimes used to produce a sense of scientific significance. While "clinical trials," controlled or not, are based on solid theoretical grounds and supported by diligently-developed hypotheses, "trial and error" is based on hunches with very limited scientific scrutiny. It is generally based on a hunch that A is good for B. To prove this, volunteers with B are recruited to receive A, in whom changes in B are measured. Next, based on the frequency and strength of these changes, A is tweaked till the desired result is observed. The problem with this approach is that it is generally not based on solid scientific grounds, it does not incorporate the possibility that changes in B may have happened anyway in the absence of A, and there is no guarantee that A always produces these changes in other circumstances.

11 — PREVENTION

Disease prevention is a concept that includes all activities aimed at a population for reducing the burden of the disease or destructive behavior and untimely death. The concept of prevention can be visualized as a pyramid in five distinct levels, each one of them covers a fraction of individuals that were not affected in the previous level of prevention. Specifically, it includes those who did not benefit from the previous level and are now moved up into an upper level with more specific and individualized actions taken at higher cost. Although complete prevention of diseases and mass events is very difficult to achieve, intelligent use of the procedure at each level can have a significant impact on the well-being of the community and reducing the economic and manpower burden of activities at the subsequent level. This concept of a tiered prevention effort can also be used for modification and increasing use of various mass phenomena in a desired direction. This is commonly used in health education and persuasive efforts toward achieving healthy nutrition or expansion of immunization. The prevention pyramid and its levels are presented in Figure 15.

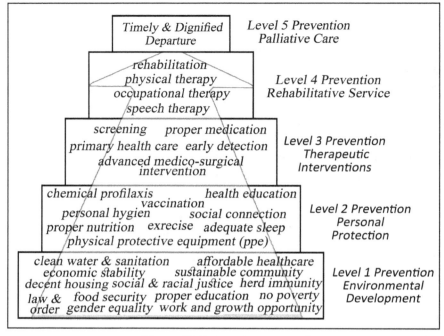

Figure 15. Prevention pyramid

Level 1. Environmental Development

The objective of this level of activities that is also called primordial prevention is to create an environment in which exposure to infectious agents or circumstances that cause the disease or the mass phenomena are eliminated or minimized. A few examples of this level of prevention are environmental sanitation and provision of clean water and sanitary waste disposal to prevent diarrhea and other waterborne diseases, maintenance and enhancement of food supply to prevent malnutrition, mass immunization and herd immunity to prevent expansion of infectious diseases, public education through mass media to enforce healthy behavior, expansion of law enforcement to prevent criminal activities, use of chemicals like fluoride to prevent tooth decay, and finally, expansion of employment and job creation to increase the economic stability of individuals and the families.

Herd Immunity

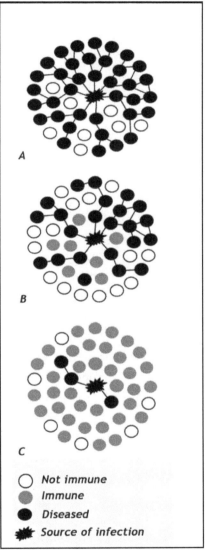

Figure 16. Herd Immunity

While immunity against a given disease is basically personal and is classified at the second level of prevention, its cumulative effect in the community, known as "herd immunity", provides a kind of prevention that can be classified as level one. Herd immunity refers to the situation when a large segment of the community is immunized and provides a barrier against propagation and transmission of the disease. Herd immunity is an important concept in prevention of infectious diseases and implies overall community protection. Technically, it is impossible to immunize all members of a given community. Some cannot be immunized because of biological resistance to vaccines, and some are not immunized because of issues surrounding various aspects of vaccination practices or even personal or political choices. If a high proportion of the population is immunized, however, the few who are not immunized, for whatever reason, are shielded against the disease. In recent years, the anti-vaxxers who refuse vaccination for a multitude of reasons, many of which has nothing to do with the disease or its biology, are enjoying the benefits of herd immunity reached by other members of the society and claim their situation as evidence against the concept of immunization. The proportion of immunized individuals in a community to achieve herd immunity is different for different diseases and is related to the speed and nature of transmission. The level of herd immunity for measles is about 95%, whereas for whooping cough, it is about 80%. Moreover, herd immunity is most effective in diseases that are directly transmitted from human to human and produce long-lasting immunity. Sexually-transmitted diseases are also limited to human transmission, but since there is no lasting immunity for them, the concept of herd immunity does not apply to them. With the increasing number of individuals and families who refuse to get vaccinated, herd immunity will be severely compromised, and it will fail to function as a protective shield for the community. This concept is presented in Figure 16. Section A represents a community with no or very low immunized individuals. In this community, disease spreads very rapidly throughout the community. Section B represents a community with a small portion of immunized individuals. Disease spreads in this community too but not as fast or as widespread as in section A. Section C represents a community with high number of immunized individuals.

This is an example of herd immunity providing cover for the few who are not or have selected not to be immunized. Apart from vaccination, another way to achieve herd immunity is through natural and widespread infection of people by the same disease. This is practical for mild and benign infections, such as *Roseola infantum* in young children, an infection that is of short duration and provides longtime immunity. For severe infectious diseases with serious consequences or death, like COVID-19, naturally-induced herd immunity is not practical or advisable. With the expansion of herd immunity, it is possible to gain long-lasting environmental protection against infectious diseases. A good example is the eradication of smallpox from the face of the earth.

Level 2. Personal Protection

This level that is also commonly known as primary prevention and includes all interventions that are based on protecting the individual host and is aimed at providing them with some sort of defense against specific disease. Clearly, the need for personal prevention arises from the failure of environmental prevention. Travelers to areas with poor basic sanitation or rampant endemic diseases are advised to follow certain protocols to protect themselves. A well-known example is chemoprophylaxis or temporary use of medication, like anti-malarial drugs when travelling to areas with high level of malaria endemicity. Vaccination, physical protection, adoption of personal protective behavior, and health education are the main elements included in this level.

Vaccination is the most significant tool in this level that produces immunity against specific infectious diseases. The concept of immunity is based on activation of the body's defense system against specific infectious agent following natural infection, i.e. getting sick, or through artificial exposure to the less virulent or non-pathogenic type of the disease agent that cannot produce disease. Vaccination is done to healthy individuals to protect them against specific infectious diseases and preventing the transmission of the disease to other people. The Advisory Committee on Immunization Practices (ACIP) in the US is the scientific organization that determines the type of vaccines for various diseases and develops specific age vaccination schedules for children, and various age categories of the adult populations in the country. There are various kinds of vaccines that are composed of the total infectious agent which has been killed or made harmless (e.g., flu and measles vaccine). Others include mRNA (CovidmRNA (COVID-19 vaccines) or bacterial toxin like tetanus. Nevertheless, they all work in the same way and train the defensive mechanisms of individual to recognize and neutralize the infectious agent if they enter the body. Vaccination, however, is not perfect and is virtually impossible to immunize every individual in a population in any given area. The reasons for this are multiple and varied. Apart from personal choice against vaccination, there are biological reasons for some individuals not to respond properly, and strategic reasons for maintaining the required transportation temperature, as well as incorrect practices by the health workers who perform vaccination. Nevertheless, a high level of vaccination will result in herd immunity and will prevent the spread of the disease in the community as part of environmental protection.

Level 3. Diagnosis and Treatment

The next level of prevention is diagnosis and treatment. It begins with screening for disease aimed at early detection prior to manifestation of obvious clinical signs and symptoms and continues to cover all the therapeutic activities, from simple office visit to complex surgical procedures. The two words "treatment" and "cure" point to therapeutic intervention with some conceptual difference. Treatment means any activity aimed at reducing the impact of the disease, providing comfort to the patient, and improving the quality of life. Cure, on the other hand, includes all the therapeutic activities aimed at complete destruction of the cause of the disease and bringing back the patient to the status prior to disease manifestation. Most infectious diseases are curable by administration of relevant antibiotics, while many of the chronic diseases, including cancers are technically not curable and can only receive treatment.

Screening and Early Detection

Screening is a procedure in which healthy populations are examined for the presence of signs and symptoms that are indicative of probable disease and separating those who have it from the rest of the population. Those who are found positive in screening tests will be examined with more specific diagnostic tests. Screening is basically a population-based procedure. Early detection, on the other hand, is focused on the individual and aims to identify them early in their disease process when they are more amenable to treatment. Physicians look for early signs of the disease during their regular routine visits and take appropriate actions. The objective of screening is to reduce a large population with small number of cases to a smaller population with larger number of probable cases and refer them for more elaborate and expensive diagnostic tests. The schematic design of screening procedure is presented in Figure 17.

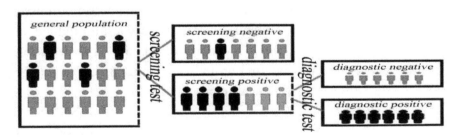

Figure 17. Disease Screening

Tests that are used for screening should be simple, easy to conduct with clear results for fast and accurate interpretation. Moreover, screening must be used only for diseases for which clear and definite therapeutic interventions are available. There is no point in identifying an individual as probably having a disease for which nothing can be done!

Table 12 presents the association of the screening test with the actual state of the disease. A+B is the total number of patients in the community, while C+D is the total of the individuals without the disease. Similarly, A+C is the total of individuals who reacted positively to the screening test and B+D are those with negative test results.

Sensitivity is the proportion of true positives (A) over the total number of diseased individuals in the community (A+B). A sensitivity of 90% means that the screening test is capable of separating 90% of all diseased individuals in the community under study among those with positive results. *Specificity* is the proportion of the true negatives (D) over the total number of not diseased individuals in the community (C+D). A specificity of 80% means that 80% of all not diseased individuals in the community are negative in the test. In other words, the positives in this screening test includes 90% of all diseased and 20% of all not diseased in the community. Screening tests are designed for use in the asymptomatic population and to funnel suspected individuals for further investigations. Screening tests are not intended to be used for diagnostic purposes, although it can be used by physicians as preliminary test for individual patients. A well-known example is mammography for early detection of breast cancer that can be used on individuals in clinical setting or mass populations is special screening programs.

Test	Disease		Total
	Present	Absent	
Positive	A	C	A+C
Negative	B	D	B+D
Total	A+B	C+D	A+B+C+D

Table 12. Possible screening test results

Two other indicators can also be calculated from data in Table 12, *Positive predictive value (PPV)* is the proportion of true cases (A) over the total number of positives in the screening test (A+C). A *PPV* of 80% means that 80% of all individuals with positive screening test are truly diseased. *Negative predictive value (NPV)* is the proportion the true negatives (D) over the number of all screening test negatives (C+D). An *NPV* of 90% means that 90% of those negative in the screening test are actually not diseased. Care must be taken not to confuse sensitivity with positive predictive value.

Level 4. Rehabilitation

The fourth level of prevention is "rehabilitation." The objective of this level as a branch of medicine is to compensate the disabling impacts of the disease by bringing back the functionality of the affected organ or replacing the lost functionality with a new one. The overall aim of rehabilitation, however, is to enable the affected individual to maintain a reasonably healthy and productive life and prevent untimely death.

Rehabilitation includes occupational, physical, and speech therapies and is highly individualized in scope and intensity. Occupational therapy provides training for regaining lost capabilities or developing new skills for engaging in the daily activities of life like dressing, eating, and so on. Physical therapy provides help with reducing pain and facilitating movements. It is given to individuals after major surgery, amputations, and other conditions with severe mobility limitation. Speech therapy is the type of rehabilitation that provides help for those with speech issues, either caused by an inherited diseases like Down's syndrome or acquired conditions such as stroke or Parkinson's disease and dementia.

Level 5. Palliative Care.

This level of prevention is "palliative" in nature and is aimed at reducing the agony and physical and mental suffering of the final days of life. The main actions at this level are intended to reduce pain and increase comfort. Do-not-resuscitate (DNR) orders are a significant part of this level of prevention. DNR orders provide caregivers with the permission to stop attempts to maintain the life of patients in the "vegetative stage," when consciousness is lost and the patient becomes irresponsive to the outside stimuli. Patients make the DNR decision while they are in full command of their mental faculties or relegate this decision to their spouse, partner, or children.

As an example of levels of prevention, consider a disease like lung cancer. Level one prevention (primordial) is achieved by enforcing no smoking in the public areas. Level two (secondary) prevention is achieved by quitting smoking. Level three (early detection) prevention is achieved by regular visits for health checks, early diagnosis, and proper treatment. Level four (rehabilitation) prevention is achieved by providing help with breathing and portable oxygen containers, and finally, level five (palliative) prevention is achieved by providing dignified hospice care.

Eradication and Elimination

Eradication is the ultimate goal in disease control. It means complete removal of the agent of the disease from the face of the earth so that no further effort would be needed to prevent or treat it. This status has already been reached for smallpox in humans. The last case of the disease was diagnosed in Somalia in 1977, and in 1980, the World Health Organization (WHO) certified smallpox as eradicated. Smallpox eradication was achieved by universal mass vaccination, vigilant monitoring and rapid diagnosis of new cases, local quarantine and treatment. Since 1980, no cases of the disease have been recorded, no vaccinations against it have been carried out, and the smallpox virus is no longer detectable anywhere in the world. This is one of the most significant human achievements in public health.[46]

Elimination is one step lower than eradication and refers to a situation where no indigenous cases of the disease are identified in a given geographical area, although the agent of the disease is present outside the area and can continue to be transmitted among people in various parts of the globe. This status is generally achieved by means of (a) environmental or primordial prevention in terms of effective and targeted interventions in the living conditions of people, (b) personal prevention in terms of mass vaccination and health education, and (c) prompt diagnosis and rapid treatment of new cases to eliminate further transmission. A few diseases are at this stage across the globe. In the United States, seven major diseases have been declared as eliminated which include poliomyelitis, diphtheria, tetanus, pertussis, mumps, measles, and rubella. All these eliminations were achieved by mass vaccination. Similar achievements have also been attained in few countries in other parts of the world. Additionally, with the help of environmental interventions, a few parasitic diseases have also been eliminated in Asian and African countries, the most significant of which are Guinea worm or dracunculiasis, River blindness or onchocerciasis, and louse and ringworm infestation. Elimination is a delicate balance of host, agent, and environment interaction, and requires vigilance and constant monitoring, including regular health education, sustained monitoring and reporting, diligent observation and enforcement of health laws and regulations, plus prompt treatment and mass immunization.

Eradication and elimination are the ultimate goals of disease prevention and control and are more easily achieved for infectious diseases which have a single and specific pathologic agent. For chronic and multifactorial diseases, elimination and eradication is a difficult task and costly to achieve.

Successful efforts for eradication and elimination of diseases require vast technical knowledge of the scientists, strong administrative will, sufficient financial support of the government, and acceptance and collaboration of the people. Lack of any of these elements can severely impair the attainment of this level of disease control.

REFERENCES

1. Holy Bible, Exodus, 7:14–24, 8:15–32, 9:1–35, 10:1–29, 11:12–36.
2. Jarus, O. 20 of the worst epidemics and pandemics in history. All About History March 20, 2020. (https://www.livescience.com/worst-epidemics-and-pandemics-in-history.html)).
3. Cukier, W.; Eagen SA, S. A. Gun violence. Curr Opin Psychol. February 2018; 19:109–112.
4. Magarey RD, R. D.; Trexler CM, C. M. Information: a missing component in understanding and mitigating social epidemics. Humanities and Social Sciences Communications. October 20, 2020;7(1):1–1.
5. White AI, A. I. Historical linkages: epidemic threat, economic risk, and xenophobia. The Lancet. April 18, 2020;395(10232):1250–1.
6. Worldometer. Covid-10 Coronavirus pandemic. https://www.worldometers.info/coronavirus/
7. Worldometer. Covid-10 Coronavirus pandemic. https://www.worldometers.info/coronavirus/
8. McMahon, B..; Pugh, TFT. F.; Ipsen, J.(1960). F. Epidemiobgic Methods.
9. Porta, M. (edt.). A Dictionary of Epidemiology, 5th Edition, Oxford University Press 2008.
10. Remington PL, P. L.; Brownson RC, R. C.; Wegner MV, M. V. Chronic Disease Epidemiology and Control, 3rd Edition. American Public Health Association, 2010.
11. Tenforde MW, M. W.; Rose EB, E. B.; Lindsell CJ, C. J.; et al. Characteristics of Adult Outpatients and Inpatients with CoviDCOVID-19 — 11 Academic Medical Centers, United States, March–May 2020. MMWR Morb Mortal Wkly Rep. ePub: June 30, 2020. doi:10.15585/mmwr.mm6926e3external icon
12. Fang, Y.; Zhang, H.; Xie, J.; Lin, M.; Ying, L.; Pang, P.; Ji, W. Sensitivity of chest CT for COVID-19: comparison to RT-PCR. Radiology. February 19, 2020:200432
13. Diabetes Control and Complications Trial Research Group. The effect of intensive treatment of diabetes on the development and progression of long-term complications in insulin-dependent diabetes mellitus. *New England Journal of medicine.Medicine.* September 30, 1993;329(14):977–86.
14. Center for Disease Control and Prevention (CDC). National Notifiable Diseases Surveillance System (NNDSS). Surveillance Case Definitions for Current and Historical Conditions (https://wwwn.cdc.gov/nndss/conditions/)
15. Domestic Violence – What Exactly Is It? (https://domesticviolence.org/)

16. Central Intelligence Agency. The CIA world factbook 2010. Skyhorse Publishing Inc.; 2009. https://www.cia.gov/the-world-factbook/field/total-fertility-rate/country-comparison
17. Life Expectancy of the World Population. https://www.worldometers.info/demographics/life-expectancy/#countries-ranked-by-life-expectancy
18. Kamineni A, Williams MA, Schwartz SM, Cook LS, Weiss NS. The incidence of gastric carcinoma in Asian migrants to the United States and their descendants. Cancer Causes Control. 1999 Feb;10(1):77-83. doi: 10.1023/a:1008849014992. PMID: 10334646.
19. Graunt, J. Natural and Political Observations Made upon the Bills of Mortality. Reprinted 1939, The Johns Hopkins Press, Baltimore.
20. Chao, F.; Gerland, P.; Cook AR, A. R.; Alkema, L. Systematic assessment of the sex ratio at birth for all countries and estimation of national imbalances and regional reference levels. Proc Natl Acad Sci USA. May 7, 2019;116(19):9303–9311. doi: 10.1073/pnas.1812593116.
21. World Health Organization Mortality Database. (https://apps.who.int/healthinfo/statistics/mortality/whodpms/).
22. Ivey-Stephenson AZ, A. Z.; Crosby AE, A. E.; Jack SPD, S. P. D.; Haileyesus, T.; Kresnow-Sedacca, M. Suicide Trends Among and Within Urbanization Levels by Sex, Race/Ethnicity, Age Group, and Mechanism of Death — United States, 2001–2015. MMWR Surveill Summ 2017;66(SS–18).
23. Tuomilehto, J.; Hu, G.; Bidel, S.; Lindström, J.; Jousilahti, P. Coffee consumption and risk of type 2 diabetes mellitus among middle-aged Finnish men and women. Jama. March 10, 2004;291(10):1213–9.
24. Brooks J. The sad and tragic life of Typhoid Mary. CMAJ: Canadian Medical Association Journal. 1996 Mar 15;154(6):915.
25. Bingham, P.; Verlander NQ, N. Q.; Cheal MJ, M. J. John Snow, William Farr and the 1849 outbreak of cholera that affected London: a reworking of the data highlights the importance of the water supply. Public Health. September 2004;118(6):387–94. doi: 10.1016/j.puhe.2004.05.007. PMID: 15313591.
26. American Cancer Society. Cancer Statistics Center. Breast Cancer Incidence 2013–2017. https://cancerstatisticscenter.cancer.org/#!/cancer-site/Breast
27. Sacks-Davis, R. et al. "The role of living context in prescription opioid injection and the associated risk of hepatitis C infection."." Addiction 111.11 (2016): 1985–1996.
28. Kittelsen, Sonja KristineS. K. et al. "Editorial: the political determinants of health inequities and universal health coverage." *Globalization and Health* vol. 15, Suppl 1 73. November 28, 2019, doi:10.1186/s12992-019-0514-6.
29. Siegel, RL., R. L.; Miller, KD., K. D.; Jemal, A. Cancer Statistics, 2020. CA Cancer J Clin 2020;70:7–30. © 2020 American Cancer Society.

30. Gramlich, John. "What the data says about gun deaths in the US, 2022" https://www.pewresearch.org/fact-tank/2022/02/03/what-the-data-says-about-gun-deaths-in-the-u-s/.
31. CDC. Drug Overdose Deaths. https://www.cdc.gov/drugoverdose/deaths/index.html
32. https://preventepidemics.org/preparedness/
33. Wilson, N.; Kariisa, M.; Seth, P.; Smith, H. IV; Davis NL, N. L. Drug and Opioid-Involved Overdose Deaths — United States, 2017–2018. MMWR Morb Mortal Wkly Rep 2020;69:290–297.
34. The World Bank: Birth Rate, Crude (per 10001,000 population). https://data.worldbank.org/indicator/SP.DYN.CBRT.IN
35. Surveillance Research Program, National Cancer Institute SEER*Stat software (www.seer.cancer.gov/seerstat) version 8.3.8.
36. Ranganathan, P.; Pramesh CS, C. S.; Buyse, M. Common pitfalls in statistical analysis: Clinical versus statistical significance. Perspect Clin Res. July–September 2015;6(3):169–70.
37. Centers for Disease Control and Prevention (CDC), National Vital Statistics System (NVSS). https://www.cdc.gov/nchs/nvss/index.htm
38. Ballesteros MF, M. F.; Williams DD, D. D.; Mack KA, K. A.; Simon TR, T. R.; Sleet DA., D. A. The Epidemiology of Unintentional and Violence-Related Injury Morbidity and Mortality among Children and Adolescents in the United States. Int. J. Environ. Res. Public Health 2018, 15(4), 616; https://doi.org/10.3390/ijerph15040616
39. Panacek EA, E. A.; Thompson CB, C. B. Sampling methods: Selecting your subjects. *Air Medical Journal*. March 1, 2007;26(2):75–8.
40. Pew Research Center, January, 2018, "For Weighting Online Opt-In Samples, What Matters Most?"
41. Tsao CW, C. W.; Vasan RS, R. S. Cohort Profile: The Framingham Heart Study (FHS): overview of milestones in cardiovascular epidemiology. International journal of epidemiology. December 1, 2015;44(6):1800–13.
42. Cameron, N. The growth and development of cohort studies. Ann Hum Biol. March 2020; 47(2):89–93. doi: 10.1080/03014460.2020.1727012. PMID: 32429754.
43. Polack FP, F. P.; Thomas SJ, S. J.; Kitchin, N. et al; C4591001 Clinical Trial Group. Safety and Efficacy of the BNT162b2 mRNA Covid-19 Vaccine. N Engl J Med. December 31, 2020;383(27):2603–2615. doi: 10.1056/NEJMoa2034577.
44. Kelsey FO. Problems raised for the FDA by the occurrence of Thalidomide embryopathy in Germany, 1960–1961. Am J Public Health Nations Health. May 1965; 55(5):703–7. doi: 10.2105/ajph.55.5.703.

45. Joint CDC and FDA Statement on Johnson & Johnson COVID-19 Vaccine. April 13, 2021. https://www.fda.gov/news-events/press-announcements/joint-cdc-and-fda-statement-johnson-johnson-covid-19-vaccine.
46. CDC. History of Smallpox. https://www.cdc.gov/smallpox/history/history.html.

INDEX

A

age adjusted rates 49
"agent" 28
agent 21, 24, 33, 36, 65, 86
agricultural 27, 29
Analytical Studies 66
arthropod vectors 33
association 26, 45, 84

C

cancer 25, 49, 91
"carrier" 28
case-control 67
causation 24, 25, 66
CDC 56, 89
chronic diseases 24
Cluster sampling 59
Cohort 64, 91
confidence 42
Control 35, 89
controlled 70
convenience sampling 59
corruption 24
Covid-19 35
CoviD-19 89
Cross sectional 63, 68
CT 89

D

descriptive studies 65, 66
Descriptive studies 65
detection 25, 56, 84
determinants 12, 26, 36, 90
diabetes 26, 89, 90
Diphtheria 23, 87

distribution 33, 36, 41, 56, 57, 59
drug overdose 65

E

electromagnetic radiation 29
Elimination 86
endemic 34
environment 21, 22, 28, 34, 36, 65, 70
epidemic 34
"Epidemic" 34
Epidemic 34
"epidemics" 28
epidemiological studies 63, 65, 66
epidemiologist 56
eradication 86
Eradication 86, 87
Experimental studies 70
extreme temperature 24, 29

F

fecal-oral 29
Framingham Heart Study 64, 91

G

Guinea worm 87

H

hazardous chemical 23
hazardous materials 27
host 21, 22, 24, 25, 27, 33, 36, 65
hygiene 29
hypothesis 66

I

iceberg 56

immunization surveys 59
incidence 32, 45
income 22, 24
international borders 35
intervention 22, 24, 25, 36, 70, 71
interventions 25, 69, 70, 71, 86

K

Kaposi sarcoma 56

L

life expectancy 25

M

malaria 27
mammography 25, 84
Margin of error 58
marital status 24
"Mary Mallon" 28
mean 41, 42
Measles 23, 35, 87
measures of central tendency 40
median 41
metazoa 23
metropolitan 29
mode 41
mortality 25, 32, 65, 90
Mortality patterns 22
multifactorial 24, 26
Multifactorial causation 24
Mumps 87

N

Negative Predictive Value 84
Non-random sampling 60

O

obesity 25
occupation 24, 25, 29
"opioid" 37

opioid 35, 37, 90

P

Pandemic 35
Pertussis 87
"pesticides" 29
pesticides 23, 29
pneumocystis pneumonia 56
Poliomyelitis 87
Positive Predictive Value 84
poverty 24
prevalence 45
prevention 70, 71, 85, 86, 87
Prevention 89
Prospective studies 63, 68
protozoa 23
public health 36, 69, 86

Q

Quota sampling 60

R

Rabies 29
race and ethnicity 24, 57
Randomized Controlled Trial 70
Random selection 57
rate 26
ratio 25
reference population 57, 58, 59
refutable 66
Rehabilitation 85
Retrospective 67
Retrospective studies 63, 67
ringworm 87
risk factor 24, 68
River blindness 87
RT-PCR 89
Rubella 87

S

sampling 56, 57, 58
Sampling frame 58
sanitation 29, 71
screening 25, 83, 84
secular trend 66
Sensitivity 84, 89
sex ratio 25, 90
sexually transmitted 28, 35
Simple random sampling 59
Small pox 86
smoking 25
snow ball sampling 60
sound blasts 29
specific criteria 12
specificity 84
Specificity 84
Sporadic 33
standard deviation 41
standard error 42
standard population 26, 49
Statistics 90
Stratified sampling 59
swine flue 29
Systematic sampling 59

T

Tetanus 87
tobacco 23
total count 56
toxins 23
transmission 21, 36, 87
treatment 34, 70, 71, 86, 87, 89
"typhoid" 28

V

vaccination 86, 87

W

WHO 86, 90

Z

"zoonotic" 29
zoonotic 22, 29

CPSIA information can be obtained
at www.ICGtesting.com
Printed in the USA
LVHW040047270523
748224LV00010B/146/J